LIVING EARTH COMMUNITY

Living Earth Community

Multiple Ways of Being and Knowing

Edited by Sam Mickey,
Mary Evelyn Tucker, and John Grim

OpenBook Publishers

https://www.openbookpublishers.com

ISBN Paperback: 978-1-78374-803-7
ISBN Hardback: 978-1-78374-804-4
ISBN Digital (PDF): 978-1-78374-805-1
ISBN Digital ebook (epub): 978-1-78374-806-8
ISBN Digital ebook (mobi): 978-1-78374-807-5
ISBN XML: 978-1-78374-808-2
DOI: 10.11647/OBP.0186

Cover image: *Feathers and Fins* (2014) by Nancy Earle, all rights reserved.
Cover design: Anna Gatti.

Contents

Fig. A1 Garden Aerial. Oak Springs Garden Foundation House, Upperville, Virgina. Photograph by Max Smith (2018), CC BY.

Acknowledgments

This book, like every other book ever written, is dependent in many ways on the living, breathing Earth. As the editors of this book, we want to acknowledge the kinship, nourishment, shelter, companionship, and inspiration provided by the living Earth community. Without that figurative and quite literal support, this book would not exist. Along with gratitude for our planetary home, we gratefully acknowledge all of those who have been part of this book project directly or indirectly.

Many thanks are owed to each of the contributors for their thoughtful engagement in this collaborative project. It was a privilege and a pleasure to facilitate the gathering of such profoundly thoughtful, sensitive, and visionary people in person, and to incorporate their contributions into a single volume. This book is based on a unique workshop that took place at the Oak Spring Garden Foundation in Virginia in October of 2018.

In this beautiful setting, between delicious meals and walks on the grounds, the participants shared their creative ideas in a synergy that was deeply felt by all. From that beautiful Virginia land, cultivated for so many decades by Bunny and Paul Mellon, these ideas took different shapes and forms in lively dialogue. Old friendships were renewed, and new friendships were formed. The land wove us into itself and held us in a place of awe and wonder.

The workshop was organized by Mary Evelyn Tucker and John Grim along with Peter Crane, their former Dean at the Yale School of Forestry and Environmental Studies and the current President of the Oak Spring Garden Foundation. As a paleobotanist, Peter has done remarkable work uncovering flower, plant, and seed fossils embedded in deep time. We were assisted at Oak Spring by Peter's wife, Elinor Crane, and especially by the dedicated preparation of program officer, Marguerite Hardin. Max Smith, the head of communications, filmed the interviews that we are posting along with this book. The staff at Oak Spring deserve our gratitude for exquisite meals and care in so many ways.

Special thanks are owed to Alessandra Tosi, Laura Rodriguez, Adèle Kreager, Luca Baffa and all those at Open Book Publishers, whose commitments to rigorous scholarly standards, service to the public good, and open access publishing fit perfectly with the spirit of this project. Gratitude to Mark Turin for the introduction.

In turn, Mary Evelyn and John would like to acknowledge the assistance of Sam Mickey on this project. Sam has been responsible for bringing the manuscript into being after the workshop, and we are enormously grateful for this. Likewise, our long-term assistant, Tara Trapani, was indispensable in the organization of the workshop, to which she brought her remarkable attention to detail. We were delighted to have Susan O'Connor with us at Oak Spring, and we further wish to acknowledge the ongoing assistance of the Charles Engelhard Foundation for our work. Similarly, we thank Nancy Klavans and the Germeshausen Foundation for their steadfast support over many years. We are also grateful to Nancy Earle for her beautiful painting of *Feathers and Fins* for the cover of the book.

Sam would like to express deep appreciation for his students and colleagues at the University of San Francisco and the California Institute of Integral Studies. He is also grateful for his collaborations with the Yale Forum on Religion and Ecology as well as the Journey of the Universe project. He is thankful for friends and family, who have given encouragement and support far beyond what can be listed here. Finally, many thanks are owed to Kimberly Carfore for sharing her love, partnership, and practice of the wild.

On behalf of the editors and all the contributors, with gratitude for all our relations, this book is dedicated to the living memory of our ancestors and evolutionary pasts, and to the future flourishing of a vibrant Earth community.

Sam Mickey
Mary Evelyn Tucker
John Grim

Fig. A2 Morning Garden. Oak Spring Garden Foundation, Upperville, Virginia.
Photograph by Max Smith (2018), CC BY.

Notes on the Contributors

David Abram

Dr. David Abram, cultural ecologist and geophilosopher, is the author of *The Spell of the Sensuous: Perception and Language in a More-Than-Human World* (1996) and *Becoming Animal: An Earthly Cosmology* (2010). Hailed as 'revolutionary' by the *Los Angeles Times*, and as 'daring' and 'truly original' by *Science*, Abram's work has been catalytic for the emergence of several new disciplines, including the steadily growing field of ecopsychology (in both its clinical and its research branches). His essays on the cultural causes and consequences of environmental disarray are published in numerous magazines, scholarly journals, and anthologies. A recipient of the international Lannan Literary Award, as well as fellowships from the Rockefeller and the Watson Foundations, in 2014 Abram held the international Arne Næss Chair in Global Justice and Ecology at the University of Oslo. Abram's work engages the ecological depths of the imagination, exploring the ways in which sensory perception, language, and wonder inform the relation between the human body and the breathing earth. His philosophical craft is informed by his fieldwork with Indigenous peoples in southeast Asia and the Americas, as well as by the European tradition of phenomenology. His ideas are often discussed and debated (sometimes heatedly) within the pages of various academic journals, including *Environmental Ethics* and the *Journal of Environmental Philosophy*.

Abram was the first contemporary philosopher to advocate for a reappraisal of 'animism' as a complexly nuanced and uniquely viable worldview, one which roots human cognition in the dynamic sentience of the body while affirming the ongoing entanglement of our bodily experience with the uncanny intelligence of other animals, each of whom encounters the same world that we perceive yet from an outrageously different angle and perspective. A close student of the Traditional Ecological Knowledges (TEK) of diverse Indigenous peoples, Abram's work also articulates the entwinement of human subjectivity with the varied sensitivities of the many plants upon which we depend, as well as with the agency and dynamism of the particular places, or bioregions, that surround and sustain our communities. In recent years, Abram's work has come to be associated with a broad movement loosely termed 'New Materialism', due to his espousal of a radically transformed sense of matter and materiality. A Distinguished Fellow of Schumacher College in England, Abram is founder and creative director of the Alliance for Wild Ethics (AWE), a consortium of individuals and organizations dedicated to cultural metamorphosis through a rejuvenation of place-based oral culture. He lives with his two children in the foothills of the southern Rockies.

Frederique Apffel-Marglin

Frédérique Apffel-Marglin, Emerita Professor in Anthropology, Smith College, founded the nonprofit organization, Sachamama Center for Biocultural Regeneration in 2009, dedicated to the regeneration of both the local forest and of indigenous agriculture and culture in the Peruvian Upper Amazon. The center is an educational organization that aims to integrate theory, research, activism, and spirituality. Apffel-Marglin was a research adviser at the World Institute for Development Economics Research (WIDER) in Helsinki, an affiliate of the United Nations University. With the Harvard economist Stephen Marglin, she formed an interdisciplinary and international collaborative team that produced three books on critical approaches to development and globalization. Her newest book, co-authored with Robert Tindall and David Shearer, is titled *Sacred Soil: Biochar and the Regeneration of the Earth* (2017). She has published an additional thirteen books, including (with Tariq Banuri) *Who Will Save the Forests?: Knowledge, Power and Environmental Destruction* (1993), and (with Stephen A. Marglin) *Decolonizing Knowledge: From Development to Dialogue; A Study Prepared for the World Institute for Development Economics Research of the United Nations University* (1996). Her interests cover ritual, gender, political ecology, critiques of development, science studies, and Andean-Amazonian shamanism. Her areas of specialization are South Asia and the Amazonian Andes. She currently resides in Cambridge, Massachusetts, as well as in Lamas, San Martin, Peru, the field campus of her nonprofit organization.

Jeannette Armstrong

Jeannette Armstrong is Syilx Okanagan, a fluent speaker and teacher of the Nsyilxcn Okanagan language and a traditional knowledge keeper of the Okanagan Nation. She is a founder of En'owkin, the Okanagan Nsyilxcn language and knowledge institution of higher learning of the Syilx Okanagan Nation. She currently is Assistant Professor and Canada Research Chair in Indigenous Okanagan Philosophy at the University of British Columbia Okanagan. She has a PhD in Environmental Ethics and Syilx Indigenous Literatures. She is the recipient of the Eco Trust Buffett Award for Indigenous Leadership and, in 2016, the BC George Woodcock Lifetime Achievement Award. She is an author whose published works include poetry, prose, and children's literary titles, and academic writing on a wide variety of Indigenous issues. She currently serves on Canada's Traditional Knowledge Subcommittee of the Committee on the Status of Endangered Wildlife in Canada.

Samara Brock

Samara Brock is pursuing her PhD at the Yale School of Forestry and Environmental Studies. She holds an MA in Community and Regional Planning from the University of British Columbia, and an MA in Food Culture from the University of Gastronomic Sciences in Italy. She has worked in international agriculture in Cuba and Argentina, as a food systems planner for the City of Vancouver, and, more recently, as a program officer for the Tides Canada Foundation, funding nonprofit organizations working on complex conservation, climate change, and food security initiatives. Her current research focuses on the development of environmental knowledge and expertise through engaging with organizations that are attempting to transform the future of the global food system.

Timothy Brown

Timothy Brown is Manager of University Initiatives for the National Geographic Society, a new venture dedicated to building partnerships with key universities through live student events in science and storytelling. A conservation biologist by training, he researched Canada lynx for the United States Fish and Wildlife Service before becoming a high school environmental science teacher. After eight years as an award-winning educator, he returned to graduate school at the Yale School of Forestry and Environmental Studies, where he studied environmental anthropology and served as Editor of *Sage Magazine*. He then served as a communications officer for the School, where he organized the Science and Storytelling Symposium in 2016. Timothy, who holds a BSc in conservation biology and a BA in music, was a founding steering committee member of the Yale Environmental Humanities Initiative. In addition to his work with *National Geographic*, he serves as Editor of *Connecticut Woodlands*, a quarterly publication of the Connecticut Forest & Park Association. He lives in New Haven, Connecticut, with his wife and son.

Paul Berne Burow

Paul Berne Burow is a PhD Candidate at the Yale University's School of Forestry and Environmental Studies and Department of Anthropology. Burow's work examines the culture, history, and politics of forests in the US Intermountain West with a particular focus on social belonging, novel ecosystems, and settler colonialism. Burow holds an MPhil in anthropology and MESc in forestry and environmental science from Yale University and a BA in economics and international relations from the University of California, Davis. Burow's work has previously appeared in *Environment & Society*, *Environmental Research Letters*, and Springer's *Studies in Human Ecology and Adaptation* series.

Michael R. Dove

Michael R. Dove is the Margaret K. Musser Professor of Social Ecology, Curator of Anthropology in the Peabody Museum of Natural History, and Professor in the Department of Anthropology, Yale University. His most recent books are *Bitter Shade: The Ecological Challenge of Human Consciousness* (2021), *Climate Cultures: Anthropological Perspectives on Climate Change* (co-edited with Jessica Barnes, 2015); *Science, Society, and Environment: Applying Physics and Anthropology to Sustainability* (co-authored with Daniel M. Kammen, 2015); and *The Anthropology of Climate Change: A Historical Reader* (editor, 2014).

Prasenjit Duara

Prasenjit Duara is the Oscar Tang Chair of East Asian Studies at Duke University. He was born and educated in India, and received his PhD in Chinese History from Harvard University. He was previously Professor and Chair of the Department of History and Chair of the Committee on Chinese Studies at the University of Chicago (1991–2008). Subsequently, he became Raffles Professor of Humanities and Director, Asia Research Institute at National University of Singapore (2008–2015). Duara is also the President of the American Association for Asian Studies (2019–2020).

In 1988, Duara published *Culture, Power and the State: Rural North China, 1900–42*, which won the Fairbank Prize of the American Historical Association (AHA) and the Levenson Prize of the Association for Asian Studies (AAS), USA. Among his other books are *Rescuing History from the Nation* (1995), *Sovereignty and Authenticity: Manchukuo and the East Asian Modern* (2003), and, most recently, *The Crisis of Global Modernity: Asian Traditions and a Sustainable Future* (2014). He has edited *Decolonization: Now and Then* (2004) and co-edited *A Companion to Global Historical Thought* with Viren Murthy and Andrew Sartori (2014). His work has been widely translated into Chinese, Japanese, Korean and the European languages. He was awarded the *doctor philosophiae honoris causa* from the University of Oslo in 2017.

Heather Eaton

Heather Eaton holds an interdisciplinary doctorate in theology, feminism, and ecology from Saint Michael's College at the University of Toronto and is a Professor in Conflict Studies at Saint Paul University in Ottawa, Canada. She works in engaging religions on ecological, social and ethical issues. She has published extensively on ecofeminism, ecospirituality, cosmology, and ecojustice, as well as the intersection of science, evolution, and religion. Her main publications are: *Advancing Nonviolence and Social Transformation: New Perspectives on Nonviolent Theories* (2016), which she co-edited with Lauren Levesque; *The Intellectual Journey of Thomas Berry: Imagining the Earth Community*, ed. (2014); *Ecological Awareness: Exploring Religion, Ethics and Aesthetics* (2011), which she co-edited with Sigurd Bergmann; *Introducing Ecofeminist Theologies* (2005); *Ecofeminism and Globalization: Exploring Religion, Culture, Context* (2003), which she co-edited with Lois Ann Lorentzen; 'Evolution', *Worldviews: Environment, Culture, Religion*, ed. (2007); 'Gender, Religion and Ecology', *Ecotheology*, ed. (2006); *Worldviews: Environment, Culture, Religion*, ed. (2001), and many additional book chapters and articles. Eaton works as a socially engaged academic with various national and international groups on religion, ecology, social issues, animal rights, nonviolence, and peace.

John Grim

John Grim is a Senior Lecturer and Research Scholar teaching in the joint MA program in religion and ecology at Yale University School of Forestry and Environmental Studies and Yale Divinity School. He is co-founder and co-director of the Forum on Religion and Ecology at Yale with his wife, Mary Evelyn Tucker. With Tucker, Grim directed a ten-conference series and book project at Harvard on 'World Religions and Ecology'. Grim is the author of *The Shaman: Patterns of Religious Healing among the Ojibway Indians* (1983), and editor of *Indigenous Traditions and Ecology: The Interbeing of Cosmology and Community* (2001). Grim and Tucker are co-authors of *Ecology and Religion* (2014), and co-editors of the following volumes: *Worldviews and Ecology* (1994); *Religion and Ecology: Can the Climate Change?* (2001); *Thomas Berry: Selected Writings on the Earth Community* (2014); and *Living Cosmology: Christian Responses to Journey of the Universe* (2016). With Willis Jenkins, they also edited the *Routledge Handbook on Religion and Ecology* (2016). Grim is the Co-Executive Producer of the Emmy Award-winning film, *Journey of the Universe* (2011). He is the President of the American Teilhard Association.

David L. Haberman

David L. Haberman is Professor and former Chair in the Department of Religious Studies at Indiana University Bloomington. He holds a PhD in History of Religions from the University of Chicago. Although he has studied and taught a great variety of religious traditions, he specializes in Hinduism and has spent many years conducting ethnographic and textual research in India. He also teaches courses on religion, ecology, and environmentalism. His recent work focuses on the worshipful interaction with natural forms of divinity in India, such as rivers, trees, stones, and mountains. Much of Haberman's work has centered on the culture of Braj, an active pilgrimage site in northern India long associated with Krishna and known for its lively temple festivals, performative traditions, and literary creations. His publications include *Acting as a Way of Salvation: A Study of Raganuga Bhakti Sadhana* (1984), *Journey Through the Twelve Forests: An Encounter with Krishna* (1994), *The Bhaktirasamritasindhu of Rupa Gosvamin* (2003), *River of Love in an Age of Pollution: The Yamuna River of Northern India* (2006), and *People Trees: Worship of Trees in Northern India* (2013). He is currently working on a new book, provisionally titled 'Loving Stones: Making the Impossible Possible in the Worship of Mount Govardhan'.

David Haskell

David Haskell's work integrates scientific, literary, and contemplative studies of the natural world. His latest book, *The Songs of Trees: Stories from Nature's Great Connectors* (2017), examines the many ways that trees and humans are connected. The book was winner of the 2020 Iris Book Award and the 2018 John Burroughs Medal. Deborah Blum (Pulitzer winner, author of *The Poisoner's Handbook* (2010), and Director of the Knight Science Journalism program at MIT) describes *The Songs of Trees* as 'compelling, lyrical, wise', and Haskell himself as perhaps 'the finest literary nature writer working today'. Haskell's first book, *The Forest Unseen: A Year's Watch in Nature* (2012), was winner of the National Academies' Best Book Award for 2013, finalist for the 2013 Pulitzer Prize in nonfiction, winner of the 2013 Reed Environmental Writing Award, winner of the 2012 National Outdoor Book Award for Natural

History Literature, runner-up for the 2013 PEN E. O. Wilson Literary Science Writing Award, and winner, in its Chinese translation, of the 2016 Dapeng Nature Writing Award. A profile by James Gorman in *The New York Times* said of Haskell that he 'thinks like a biologist, writes like a poet, and gives the natural world the kind of open-minded attention one expects from a Zen monk rather than a hypothesis-driven scientist'. E. O. Wilson wrote that *The Forest Unseen* was 'a new genre of nature writing, located between science and poetry'. *The Forest Unseen* has been translated into ten languages.

Haskell has also written about the biology of climate change, and same-sex marriage for *The New York Times*. Haskell holds degrees from the University of Oxford (BA) and from Cornell University (PhD). He is Professor of Biology and Environmental Studies at the University of the South, where he served as Chair of Biology. He is a 2014–2015 Fellow of the John Simon Guggenheim Memorial Foundation, a Fellow of the American Council of Learned Societies, and an Elective Member of the American Ornithologists' Union. His scientific research on animal ecology, evolution, and conservation has been sponsored by the National Science Foundation, the Environmental Protection Agency, the Fish and Wildlife Service, and the World Wildlife Fund, among others. He serves on the boards and advisory committees of local and national land conservation groups. Haskell's classes have received national attention for the innovative ways they combine action in the community with contemplative practice. In 2009, the Carnegie and Case Foundations named him Professor of the Year for Tennessee, an award given to college professors who have achieved national distinction and whose work shows 'extraordinary dedication to undergraduate teaching'. In 2011, the *Oxford American* featured him as one of the southern US's most creative teachers. His teaching has been profiled in *USA Today*, *The Tennessean*, and other newspapers.

Willis Jenkins

Willis Jenkins lives in the Rivanna River watershed, Virginia, where he works as Professor of Religion, Ethics, and Environment at the University of Virginia. With Matthew Burtner, Jenkins leads the Coastal Futures Conservatory, a transdisciplinary initiative focused on listening to environmental change using different ways of knowing drawn from the arts, humanities, and sciences. With Mary Evelyn Tucker and John Grim, he is co-editor of the *Routledge Handbook of Religion and Ecology* (2016). He is author of two award-winning books: *Ecologies of Grace: Environmental Ethics and Christian Theology* (2008), which won a Templeton Award for Theological Promise, and *The Future of Ethics: Sustainability, Social Justice, and Religious Creativity* (2013), which won an American Academy of Religion Award for Excellence in the Study of Religion.

Sean Kelly

Sean Kelly, PhD, is professor of Philosophy, Cosmology, and Consciousness at the California Institute of Integral Studies (CIIS). He is the author of *Coming Home: The Birth and Transformation of the Planetary Era* (2010), co-editor of *The Variety of Integral Ecologies: Nature, Culture, and Knowledge in the Planetary Era* (2017), and co-translator of Edgar Morin's *Homeland Earth: A Manifesto for the New Millennium* (1999). Along with his academic work, Kelly teaches taiji and is a facilitator of the group process Work that Reconnects developed by Joanna Macy.

Eduardo Kohn

Eduardo Kohn is Associate Professor of Anthropology at McGill University. He is best known for the book, *How Forests Think: Toward an Anthropology beyond the Human* (2013), which has been translated into several languages. It won the 2014 Gregory Bateson Prize. His research continues to be concerned with capacitating sylvan thinking in its many forms as a means of fashioning an ethics for living on a planet in ecological crisis.

Thomas E. Lovejoy

Thomas E. Lovejoy was elected University Professor at George Mason in March 2010. He previously held the Biodiversity Chair at the Heinz Center for Science, Economics and the Environment and was President from 2002–2008. An ecologist who has worked in the Brazilian Amazon since 1965, he works on the interface of science and environmental policy. Starting in the 1970s, he helped bring attention to the issue of tropical deforestation, and, in 1980, published the first estimate of global extinction rates (in *The Global 2000 Report to the President*). He conceived the Minimum Critical Size of Ecosystems project (also known as the Biological Dynamics of Forest Fragments Project) — a long-term study on forest fragmentation in the Amazon (which began in 1978), and which is the largest experiment in landscape ecology to date. He also coined the term 'Biological diversity', originated the concept of debt-for-nature swaps, and has worked on the interaction between climate change and biodiversity for more than thirty years. He is the founder of the public television series *Nature*. In the past, he served as the Senior Advisor to the President of the United Nations Foundation, as the Chief Biodiversity Advisor to the World Bank as well as Lead Specialist for the Environment for the Latin American region, as the Assistant Secretary for Environmental and External Affairs for the Smithsonian Institution, and as Executive Vice President of World Wildlife Fund-US. In 2002, he was awarded the Tyler Prize, and, in 2009, he was the winner of BBVA Foundation Frontiers of Knowledge Award in the Ecology and Conservation Biology Category. In 2012, he received the Blue Planet Prize. He has served on advisory councils in the Reagan, George H. W. Bush, and Clinton administrations. In 2009, he was appointed Conservation Fellow by the National Geographic Society. He chaired the Scientific and Technical Panel for the Global Environment Facility which provides funding related to the international environmental conventions from 2009–2013 and serves as Advisor to the current Chair. He received his BSc and PhD (Biology) from Yale University.

Sam Mickey

Sam Mickey, PhD, is an Adjunct Professor in the Theology and Religious Studies department and the Environmental Studies program at the University of San Francisco. He has worked for several years at the Forum on Religion and Ecology at Yale. His teaching, writing, and research are oriented around the ethics and ontologies of nonhumans, and the intersection of religious, scientific, and philosophical perspectives on human-Earth relations. He is an author of several books, including *Whole Earth Thinking and Planetary Coexistence* (2015), *Coexistentialism and the Unbearable Intimacy of Ecological Emergency* (2016), and *On the Verge of a Planetary Civilization: A Philosophy of Integral Ecology* (2014). He is co-editor (with Sean Kelly and Adam Robbert) of *The Variety of Integral Ecologies: Nature, Culture, and Knowledge in the Planetary Era* (2017).

Mitchell Thomashow

Mitchell Thomashow devotes his life and work to promoting ecological awareness, sustainable living, creative learning, improvisational thinking, constructive networking, and organizational excellence. Currently, he is engaged in teaching, writing, and executive consulting, cultivating opportunities and exchanges that transform how people engage with ecological learning, sustainability, and the arts. He is the author of three books: *Ecological Identity* (1995), *Bringing the Biosphere Home* (2001), and *The Nine Elements of a Sustainable Campus* (2014). His new book, *To Know the World: Why Environmental Learning Matters*, will be published by The MIT Press in 2020. From 2015–2017, he served as the Sustainability Catalyst Fellow at Philanthropy Northwest in Seattle, Washington. In 2017, Thomashow wrote a report, *Pacific Northwest Changemakers*, that profiles eight exemplary community-based, grassroots sustainability projects in both rural and urban regions. From 2011–2015, he was the Director of the Second Nature Presidential Fellows Program, working with university presidents to promote a comprehensive sustainability agenda on their campuses. From 2006–2011, he was the President of Unity College, Maine. With his management team, he integrated concepts of ecology, sustainability, natural history, wellness, participatory governance, and community service into all aspects of college and community life. From 1976–2006, he was the Chair of the Environmental Studies Department at Antioch University New England. He founded an interdisciplinary environmental studies doctoral program and worked collaboratively to grow and nourish a suite of engaging Masters programs, geared towards working adults. Thomashow lives in the hill country of southwest New Hampshire in the shadow of Mount Monadnock. He loves to explore the fields, forests, wetlands, hills, and lakes of Northern New England.

His recreational interests include basketball, baseball, bicycling, board games, jazz piano, electronic keyboards, guitars, hiking, and lake swimming.

Mary Evelyn Tucker

Mary Evelyn Tucker is co-director with John Grim of the Forum on Religion and Ecology at Yale. Her special area of study is Asian religions. She received her PhD from Columbia University in Japanese Confucianism. Since 1997, she has been a Research Associate at the Reischauer Institute of Japanese Studies at Harvard. Her Confucian publications include: *Moral and Spiritual Cultivation in Japanese Neo-Confucianism* (1989) and *The Philosophy of Qi* (2007). With Tu Weiming, she edited two volumes on *Confucian Spirituality* (2003, 2004). Her concern for the growing environmental crisis, especially in Asia, led her to organize with John Grim a series of ten conferences on World Religions and Ecology at the Center for the Study of World Religions at Harvard (1995–1998). Together they are series editors for the ten volumes from the conferences distributed by Harvard University Press. In this series she co-edited *Buddhism and Ecology* (1997), *Confucianism and Ecology* (1998), and *Hinduism and Ecology* (2000). Tucker and Grim wrote *Ecology and Religion* (2014) and, with Willis Jenkins, edited the *Routledge Handbook on Religion and Ecology* (2016). With Brian Thomas Swimme she wrote *Journey of the Universe* (2011) and, with John Grim, was Executive Producer of the Emmy Award-winning *Journey of the Universe* (2011) film.

Mark Turin

Mark Turin, PhD, is an anthropologist, linguist, and occasional radio presenter, and an Associate Professor of Anthropology at the University of British Columbia. From 2014–2018, Turin served as Chair of the First Nations and Endangered Languages Program and from 2016–2018, as Acting Co-Director of the University's new Institute for Critical Indigenous Studies. Before joining UBC, Turin was an Associate Research Scientist with the South Asian Studies Council at Yale University, and the Founding Program Director of the Yale

Himalaya Initiative. Turin directs both the World Oral Literature Project, an urgent global initiative to document and make accessible endangered oral literatures before they disappear without record, and the Digital Himalaya Project, which he Co-Founded in 2000 as a platform to make multi-media resources from the Himalayan region widely available online. For over twenty years, Turin's regional focus has been the Himalayan region (particularly Nepal, northern India and Bhutan), and, more recently, the Pacific Northwest. Turin is very privileged to have had the opportunity to work in collaborative partnership with members of the Thangmi-speaking communities of eastern Nepal and Darjeeling district in India since 1996, and, since 2014, with members of the Heiltsuk First Nation through a Language Mobilization Partnership in which UBC is a member. He is the author or co-author of four books, three travel guides, the editor of twelve volumes, and the editor of a series on oral literature.

Paul Waldau

Paul Waldau is an educator who works at the intersection of animal studies, law, ethics, religion, and cultural studies. A Professor at Canisius College in Buffalo, New York, Waldau has been the senior faculty member for the Master of Science graduate program in Anthrozoology since its founding in 2011, and was the Program Director from 2014–2017. Waldau also taught Animal Law at Harvard Law School from 2002 to 2014, and he has been teaching courses since 2009 in Harvard's Summer School, through which he offered the online course 'The Animal-Human Divide' in Summer 2018. Waldau was one of the organizing members of the Great Ape Project, and served from 1995 through 2008 in the capacity of board member, as well as serving for years as the Vice-President and Executive Director. After helping to found the Animals and Religion group at the American Academy of Religion, Waldau spent a decade teaching ethics and public policy at Tufts University School of Veterinary Medicine, where he also was the Director of the Center for Animals and Public Policy until 2009. He has completed five books, the most recent of which are *Animal Studies — An Introduction* (2013) and *Animal Rights* (2011). He is also co-editor of the groundbreaking *A Communion of Subjects: Animals in Religion, Science, and Ethics* (2006). His first book was *The Specter of Speciesism: Buddhist and Christian Views of Animals* (2001).

Julianne Lutz Warren

Julianne Warren is a freelance storyteller and community co-organizer with an MA in linguistics and a PhD in ecology. She authored an intellectual biography, *Aldo Leopold's Odyssey* (2006, 2016), developing this influential conservationist's 'land health' concept—one that deeply resists industrial-capitalist assumptions, yet not settler colonial and racist ones. Her current personal-professional work takes up decolonization and anti-racism, and supports liberation of Native peoples (for ALL the people). Her creative pieces, including sound arts, continue exploring what it is to be in good relations. These appear in a variety of venues including, *Newfound, Minding Nature, Zoomorphic, The Poetry Lab* of The Merwin Conservancy, *Lost and Found Theatrum Anatomicum*, and The Deutsches Museum. As a faculty member, Julianne collaborated with students growing NYU Divest: Go Fossil Free! While living far north, she served as a council member of Fairbanks Climate Action Coalition and co-facilitator of #KeepItIntheGround! working group. She is a named Scholar and Fellow at the Center for Humans and Nature and an Ecosphere Studies collaborator and visiting scholar at The Land Institute. She currently inhabits ancestral Tewa homelands of Northern New Mexico.

Brooke Williams

Brooke Williams has spent the last thirty years advocating for wilderness. He is the author of four books, including *Open Midnight: Where Ancestors and Wilderness Meet* (2017), *Halflives: Reconciling Work and Wildness* (1999), and *The Story of My Heart* (2014), by Richard Jeffries, as rediscovered by Brooke Williams and Terry Tempest Williams. His journalistic pieces have appeared in *Outside, The Huffington Post, Orion*, and *Saltfront*. He and his wife, Terry Tempest Williams, divide their time between Utah and Cambridge, MA.

Preface

Sam Mickey

There are many ways of seeing Earth. It is possible to gaze at the planet from the vantage of a space shuttle in orbit. If you are standing on the moon, you can see Earth rise in the distance, as seen in the famous photograph of Earth taken from the moon by the NASA astronaut William Anders in 1968, *Earthrise* (see Figure 1). You can also look at Earth much more closely, on a more minute level, observing the habitats and inhabitants of Earth as they appear at any moment, and in any context — urban, rural, or wild. You are looking at Earth when you see a meadow, a forest, a tree, a cat, a farm, a house, or the ground beneath your feet. Along with these different ways of visually perceiving Earth, there are also many ways of understanding Earth, spanning various fields of scientific research, the religious traditions of the world, and philosophical theories of nature. There is great diversity in how we can relate to the vast panoply of beings composing the life, land, air, and water of Earth. This book is a celebration and revitalization of that diversity.

Everything lends itself to multiple perspectives. Consider the heart: an organ that is found in fish, reptiles, birds, and mammals. What is the heart? There are different ways of responding to that question. A poet speaks about the heart in terms of love and loss. A biologist speaks about the heart in terms of the cardiovascular system and blood pressure. It is not that one person is right and the other wrong. Those different ways of speaking reflect different perspectives, different ways of understanding and experiencing the heart, and different ways of knowing and being in relation to the heart. A poet and a biologist can both be right. They do not have to be mutually exclusive. They can each be true at the same time. Indeed, those perspectives can be held by the same person. Each human being contains various capacities for taking different perspectives on the

   https://doi.org/10.11647/OBP.0186.21

Fig. 1 *Earthrise*. Photo by William Anders (1968), Wikimedia, public domain, https://commons.wikimedia.org/wiki/File:NASA_Earthrise_AS08-14-2383_Apollo_8_1968-12-24.jpg

world: logical, poetic, verbal, emotional, perceptual, intellectual, social, and more.

One can cultivate the artistic perspective of a poet or painter, the mathematical and logical perspective of a chemist or biologist, the verbal skills of a speech writer, the emotional intelligence of a sensitive friend, and the embodied or somatic knowledge of a swimmer or basket weaver. Furthermore, different perspectives are variously cultivated throughout human cultures and traditions. Biology, Buddhism, Hinduism, Indigenous lifeways, mathematics, and music all involve different ways of thinking, feeling, and acting. Understanding these differences is a way of understanding ourselves collectively, of understanding humankind. Furthermore, different perspectives are not taken up only by humans, but by all kinds of living beings.

Different forms of agency, sentience, significance (semiosis), intelligence, and communication are exhibited throughout the community of life on Earth. For example, research in microbiology suggests that communication takes place between bacteria, specifically through exchanges of pulses of electrical energy.[1] Communication enables bacteria to sustain themselves in communities, without which an individual bacterium cannot survive. Regarding the increasing number of scientific studies that find evidence of intelligence across all forms of life, the botanist Robin Wall Kimmerer makes the following observation during an interview with Krista Tippett: 'I can't think of a single scientific study in the last few decades that has demonstrated that plants or animals are dumber than we think. It's always the opposite, right? What we're revealing is the fact that they have extraordinary capacities [...] we're at the edge of a wonderful revolution in really understanding the sentience of other beings.'[2]

Approaching the middle of the twenty-first century, humans are learning more and more about the extraordinary capacities of life on Earth, and, at the same time, life on Earth is undergoing a profoundly troubling transformation due to the massive overexploitation and overconsumption of resources by developed (industrialized) nations. During the current period of environmental change, immensely complex challenges are facing life on Earth, including pollution, deforestation, water scarcity, climate change, and mass extinction. Such challenges cannot be sufficiently addressed through a single perspective alone. What is needed is dialogue and integration among diverse perspectives. Planetary problems call for globally coordinated responses. The inclusion of multiple ways of being and knowing is crucial for coordinating viable responses to the intensifying ecological crises occurring around the world. This anthology is a contribution toward that effort, presenting succinct essays that explore the diverse ways in which humans think, feel, and act in relation to the community of life on Earth.

Dialogue across Perspectives

The essays in this volume illuminate different ways of being in the world and the different kinds of knowledge that they entail, such as the

---

1   Gabriel Popkin, 'Bacteria Use Brainlike Bursts of Electricity to Communicate', *Quanta Magazine*, September 5, 2017, https://www.quantamagazine.org/bacteria-use-brainlike-bursts-of-electricity-to-communicate-20170905

2   Krista Tippett, 'Robin Wall Kimmerer: The Intelligence in All Kinds of Life', *On Being*, February 25 2016, https://onbeing.org/programs/robin-wall-kimmerer-the-intelligence-in-all-kinds-of-life-jul2018

traditional ecological knowledge (TEK) of Indigenous communities, the affective knowledge that comes with religious love and devotion, the scientific knowledge of a biologist, the aesthetic knowledge of someone listening to or composing music, and the embodied knowledge communicated through storytelling. It is important to emphasize that different ways of knowing are not always harmonious or even compatible. Consider an example between different religions. Some ways of knowing are oriented around belief in God, as in the monotheism of Christianity or Islam, whereas other ways of knowing suspend belief in God (i.e., agnosticism) or they explicitly believe that there is no God (i.e., atheism). In astronomy, the idea that Earth revolves around the sun (heliocentrism) is strictly incompatible with the ancient model of geocentrism, which assumed that the sun revolved around the motionless Earth. There will always be contrasts and contradictions between perspectives, especially when considering the community of life on Earth in all its diversity. The question is how to sustain a flourishing coexistence amid this radical diversity.

Integrating multiple perspectives does not mean that everyone will agree about everything all the time. It means, rather, that there is an ongoing dialogue between those perspectives, seeking shared understanding and common interests, while accepting differences. However, not all perspectives should be included in a thriving planetary civilization. Perspectives oriented around violent control, domination, or hate cannot be included in any kind of integrative dialogue, since those perspectives refuse to participate. Respectful or hospitable relations are required for dialogue to take place. If you cannot acknowledge some truth or intrinsic value in your interlocutor, then you cannot have a dialogue. Authoritarianism, racism, religious fundamentalism, and colonialism are examples of perspectives that are not amenable to dialogue. Even the perspective of a poet can become too narrow-minded to hold itself open to dialogue. A poet and a physicist cannot have a dialogue about an ocean if the poet refuses to acknowledge that there is some validity to physics (e.g., tides are caused by the gravitational pull of the moon), or if the physicist refuses to acknowledge that there is some truth in poetry (e.g., tides are the ocean's dream of the moon).

Dialogue between multiple perspectives is not only about knowledge. Ways of knowing (epistemology) are always connected to ways of being (ontology). To put it simply, epistemology implies the existence of knowers. An artist has a way of life, a way of perceiving and acting in the world, a way of being, of which an artistic way of knowing is an integral part. Ways of knowing are not merely abstract frameworks or belief systems. Frameworks and beliefs are involved with knowledge, to be sure, but knowledge only makes sense in some kind of existential context. Knowing is therefore entangled with encountering, feeling, imagining, experiencing, relating, sensing, intending, and so on. Your understanding of the world shapes and is shaped by who you are, including your opinions and beliefs, as well as the practical, emotional, embodied, historical, and material dimensions of your existence. What a gorilla knows is part of what it is like to be a gorilla. What a scientist knows is part of what it is like to be a scientist. What a rabbi knows is part of what it is like to be a rabbi. The contributors to this volume are mindful of this connection between knowing and being. Bringing together scholars, writers, and educators across the sciences and humanities, this anthology provides informative and inspiring accounts of perspectives that attend to ways of being and knowing that intimately intertwine humans with the vibrant vitality of the Earth community.

Multicultural and interdisciplinary in scope, this anthology engages with diverse cultures and traditions around the world, and draws upon academic disciplines across the sciences and humanities. It is unique for its mixture of expertise and accessibility. The authors included in this book are leading figures in their respective arenas, and in the chapters that follow they introduce contemporary research, traditional knowledge, and emerging modes of thought in ways that are accessible to the general reader while also relevant to specialists. The essays included in this volume are revised versions of what the contributors presented to one another when they met for a workshop held in the fall of 2018 at the Oak Spring Garden Foundation in Virginia. The workshop was hosted by the renowned botanist and evolutionary plant scientist Peter Crane and organized by the two directors of the Forum on Religion and Ecology at Yale, Mary Evelyn Tucker and John Grim.

The focus of the workshop, like that of this book, is the integration of multiple perspectives on the community of life on Earth.

When bringing multiple perspectives into dialogue, there is no perspective that is assumed to be the best or truest. No single perspective is given priority over the others. Any contrasts, conflicts, and comparisons between them emerge through mutual understanding and not one-sided evaluation. It is the dialogue that is given priority, the ongoing struggle for mutual understanding. Furthermore, if no single perspective is given priority, that includes the perspective of this introductory overview. The birds-eye view is not privileged over a close-up. A more general or universal perspective is not given priority over more specific, local perspectives, and vice versa. Each perspective can be understood on its own terms. Each of this book's chapters do just that: they elucidate different ways of being and knowing on their own terms, based on their own place within the evolving community of life on Earth. The aim of the book is not to determine once and for all which perspective comes out on top, but to figure out ways to move forward together.

Chapters

The chapters of this book are grouped into six sections, which reflect the diverse histories and futures of humankind in intimate relationship with the more-than-human world. The focus of Section I is precisely the presence of that which is more-than-human. Reflecting on species dynamics within the planetary biosphere, David Abram, in Chapter 1, suggests that new insight into the astonishing navigational feats of migratory animals can be gleaned by recognizing the broad Earth as a dynamic, agential player in these migrations. The long-distance movements of various animals can readily be understood as metabolic processes within the body of the living planet, not unlike the rhythmic systole and diastole of a heartbeat.

Remembered songs of extinct wattlebirds, endemic to Aotearoa New Zealand, catalyze Julianne Warren's storytelling. In Chapter 2, she spins a path from first listening to a Pākehā-narrated recording of an elder Māori performing traditional mimicry of Huia. Replaying these dead bird-human voices interacting with sounds in the near-Arctic helps her begin learning, in poet W. S. Merwin's words, to 'hear what never/ Has fallen silent.'[3] Between antipodes, ancestral echoes escape from machines, and sleeping languages live on—in loss—spellbinding companionships of hope's sound.

In Chapter 3, Paul Waldau considers possibilities for transforming human institutions (e.g., law, education, ethics, and religion) in ways that promote a flourishing Earth community. The author considers how self-actualization for humans can be found not through the arrogance of human exceptionalism, but through different expressions of humility and through a recognition of the animality of humankind.

Section II brings attention to the dynamics of forests in Latin America. Drawing on his ethnographic research among Indigenous communities in Ecuador, Eduardo Kohn considers the political and ethical implications of thinking with forests in Chapter 4. It is a diplomatic undertaking that seeks to integrate multiple ways of understanding the cosmos, and it is an ontological undertaking that rethinks the very nature of existence by recognizing the intelligence inherent in all life.

Frédérique Apffel-Marglin advocates for integral ecological healing in Chapter 5, particularly by attending to the practices of Indigenous Amazonian communities. The use of psychedelic plant medicines in Amazonian shamanism exemplifies the kind of non-rational ways of knowing that expand human consciousness beyond the individual ego and into intimate communion with the more-than-human world.

In Chapter 6, Thomas E. Lovejoy elaborates on the importance of biodiversity for the Earth community and the role of biologists therein. Bringing science together with ethical and political issues, Lovejoy articulates the responsibilities of biologists and other scientists for promoting biodiversity and addressing contemporary ecological crises.

The ecological implications of Asian traditions provide the guiding thread for the next section, Section III. In Chapter 7, Prasenjit Duara thinks with the circulating waters of oceans to articulate the complex confluence of human and natural histories, particularly with reference to Asian contexts. Whereas the fragmentation of human and natural histories contributes to ethical and political failures to address environmental issues, Duara's oceanic metaphor demonstrates how

---

3   W. S. Merwin, 'Learning a Dead Language', in *Migration: New and Selected Poems* (Port Townsend, WA: Copper Canyon Press, 2005), p. 41.

human history, including the study of history (i.e., historiography), overlaps with natural history, while these histories nonetheless operate on different temporal scales.

Religion and ecology in Hinduism is the focus of Chapter 8, with David L. Haberman elucidating the value of love and devotion as ways of connecting to the natural world. In contrast to the detachment that characterizes abstractly intellectual forms of knowledge, these ways of connecting to nature yield emotional or affective knowledge, which promotes care for the beauty and vulnerability of the natural world.

In Chapter 9, Mary Evelyn Tucker presents contributions to ecological ethics in Confucianism, highlighting the importance of Confucian cosmology for understanding the material world as vibrant and lively, not passive and inert. Confucianism facilitates an approach to ethics for which personal and social concerns are embedded in the Earth community and the whole cosmos, such that ecological concern is not separate from the practice of self-cultivation.

Section IV integrates perspectives from ecology and the humanities, with a view toward storytelling. To build a bridge between scientific and ethical perspectives on ecological issues, David Haskell advocates in Chapter 10 for contemplative exercise, in the sense of repeated, open-ended attention. Contemplative participation within the community of life deepens one's sense of ecological aesthetics, and such appreciation for the beauty of nature provides an integrative ground for ethical actions informed by scientific knowledge.

In the next chapter, advocating for the cultivation of storytelling skills, Timothy Brown shares his experience bringing science and storytelling to students, specifically through work with National Geographic. Stories provide a framework for communicating scientific information to non-specialists, for thinking across different academic disciplines, and for motivating action.

Chapter 12 attends to the role that listening plays in attuning humans to the stories of the natural world, specifically in terms of a project involving Long-Term Ecological Research oriented around conserving coastal ecosystems. Willis Jenkins describes The Conservatory Project, which integrates perspectives on environmental change from sciences, humanities, and the arts, designing ecoacoustic listening exercises that afford participants an aural sense of their embodiment and embeddedness in a dynamic environment.

Listening can facilitate a contemplative awareness that is conducive to nonanthropocentric ways of being in the world. Brooke Williams, in Chapter 13, presents a series of reflections on the conference that gave rise to this volume. Williams discusses an exercise for engaging with ecology through the imagination. Participants are guided through an imaginal encounter with ancestors, the different kinds of gifts they might bring, and the paths those gifts can be taken.

In Section V, attention is given to the resilient relationships cultivated within Indigenous lands. Chapter 14 introduces the worldview of the Okanagan people, an Indigenous people inhabiting the northwest of North America. Jeannette Armstrong describes her personal background and experience growing up as a member of the Okanagan community in the Okanagan Valley in British Columbia, Canada. She highlights the importance of intimacy with the land, taking responsibility for relationships, and building resilient communities in the face of cultural and environmental destruction.

In the next chapter, drawing attention to the contemporary resurgence of Indigenous languages, Mark Turin describes the collaborative work of linguistic and cultural revitalization in response to the destruction of Indigenous communities in settler colonial nations. While recuperating the vitality of languages, this process also facilitates the recuperation of the well-being of Indigenous communities as well as the lands within which those languages and communities are embedded.

Chapter 16 draws on the wisdom of Indigenous traditions and the world's religions, as John Grim proposes a triad of sensing, minding, and creating, to help us understand the world without separating nature from culture. All things exhibit capacities for external interaction (sensing) and an inner patterning or consciousness (minding), and those external and internal facets change over time as novel conditions arise (creating). The emergence of life from matter and of humans from other life forms can be understood as an explication of the dynamics of sensing, minding, and creating inherent in the universe.

The following chapter indicates that revitalizing Indigenous communities requires more than a recognition of tribal sovereignty. Samara Brock shows how it also requires a recuperation of Indigenous understandings of existence and ways of being. The inclusion of multiple ontologies opens up possibilities for creating relational, hybrid forms of practices that cultivate mutuality and reciprocity between humans and the land.

The final section, Section VI, concentrates on the planetary and cosmic dimension of human existence. In Chapter 18, Sean Kelly proposes that the current cultural and ecological transformations taking place on Earth are evidence of a Second Axial Age. The period between the eighth and third centuries BCE is known as the 'Axial Age', which saw the beginnings of philosophy, science, mathematics, and many of the world's religious traditions. Whereas Axial Age values were oriented around transcendent or cosmological principles (e.g., Truth, God, Oneness), values of the Second Axial Age compel humans to reorient civilization around the living Earth community — Gaia.

The next chapter reflects on the enduring quest of human beings to inhabit and understand the universe. Weaving together an account of the exterior (objective) and interior (subjective) facets of the cosmos, Heather Eaton finds the unique qualities of human subjectivity in symbolic consciousness and in the worldviews, narratives, and other systems of symbols through which humans interpret and respond to their surroundings. Along with symbols and narratives, learning about ecology involves attention to systems and interrelationships at multiple scales, from ecosystems to the biosphere.

To facilitate the cultivation of ecological imagination and promote environmental awareness, Mitchell Thomashow's concluding chapter presents proposes five qualities of environmental learning (observation, information, interpretation, expression, and manifestation). Those educational qualities are pathways for integrated ways of knowing and being in the living Earth community.

As the concluding chapters of this volume indicate, the end of the book is not the end of the journey. This whole book is a beginning, an opening for people who seek different ways to partake in planetary coexistence. In other words, this book is an invitation to new beginnings, new possibilities for living, learning, connecting, and communicating with other humans and with the more-than-human world. This includes new opportunities for the revitalization of Indigenous lands and languages; for the rejuvenation of ancient wisdom; for the inclusion of rational, emotional, embodied, animal, and ecological ways of knowing; and for the integration of humankind within a living Earth community.

What would it look like if more people became more aware of and sensitive to their relationship with the living Earth community? How would the education, government, economy, and media change? How would individuals think, feel, and act differently? The responses to those questions will vary from place to place, depending on different cultural values, and from person to person, depending on different experiences, moods, and personal backgrounds. This book does not present a framework that the reader should apply. If a framework or model is like a map, this book can be thought of more like a compass. A map is something distinctly separate from the territory that it describes and separate from the person using the map; a compass has a more participatory relationship to the territory and to the position of the person wielding the compass. The needle of a compass is composed of steel, an alloy of iron, which is responsive to Earth's electromagnetic field. A compass needle moves according to the specific place of the person using it. With compass in hand, where will you go?

As humans shift toward a more sustainable way of inhabiting the community of life on Earth, every single human being will participate in that shift differently. Each of us will navigate several overlapping concerns, including oneself along with family, friends, and strangers, whether human or more-than-human. Some might start new nonprofits or nongovernmental organizations (NGOs), whose mission is to restore ecosystems or advocate for peace and justice. Some will advocate for environmental issues in their respective sphere of influence — at home, in school, on social media, or in the workplace. Some will feel more empathy toward a companion animal, maybe a dog or cat, whose personality makes it impossible to hold up a rigid boundary that would separate humans from our nonhuman kin. Some will make changes in dietary preferences and other personal behaviors to adapt to the precarious conditions of life on Earth. Some will have conversations with friends and family. Some will have conversations with trees, listening to what trees might have to say to an inquiring mind, like one of the characters in *The Overstory* (2018), a novel by Richard Powers about the many ways humans and trees relate to one another. Sitting on the ground, leaning against a pine tree, a woman listens to what the tree is saying, 'in words before words', and then she hears it say, 'Sun and water are questions endlessly worth answering'.[4]

We all move forward in our own way, depending on the unique circumstances of our lives. There is more than one way to read this book, as there is more than one way to inhabit this planet and answer the ongoing questions of sun and water. We are making ourselves at home on Earth, and we are learning how to do so together.

Bibliography

Merwin, W. S., 'Learning a Dead Language', in *Migration: New and Selected Poems*, W. S. Merwin (Port Townsend, WA: Copper Canyon Press, 2005), p. 41.

Popkin, Gabriel, 'Bacteria Use Brainlike Bursts of Electricity to Communicate', *Quanta Magazine*, September 5 2017, https://www.quantamagazine.org/bacteria-use-brainlike-bursts-of-electricity-to-communicate-20170905

Powers, Richard, *The Overstory: A Novel* (New York, NY: W. W. Norton and Company, 2018).

Tippett, Krista, 'Robin Wall Kimmerer: The Intelligence in All Kinds of Life', *On Being*, February 25 2016, https://onbeing.org/programs/robin-wall-kimmerer-the-intelligence-in-all-kinds-of-life-jul2018

---

4   Richard Powers, *The Overstory: A Novel* (New York, NY: W. W. Norton and Company, 2018), p. 3.

Introduction

Ways of Knowing,
Ways of Valuing Nature

John Grim and Mary Evelyn Tucker

The contemporary market-driven worldview relies upon and legitimates rational, analytical ways of knowing, often to the exclusion of other ways of knowing. Support for a consumerist ideology depends upon and simultaneously contributes to a worldview based on the instrumental rationality of the human. In this worldview, rational choice is seen as that realm of common sense in which both the world and human demands on the world are laid out as commensurate, equal realities that confront decision-makers. That is, in this rational scheme, the assumption for decision-making is that all choices are equally clear and measurable. According to that perspective, the challenge is to find a common metric for evaluating the quantitative differences among the relevant factors. Different values are integrated into this metric by assuming that all values are relative and that trade-offs are made between these values in order to arrive at a choice.

The metrics used may vary, but in the current market-driven worldview, metrics such as price, utility, and efficiency dominate. This can result in highly diverse views of a forest, for example, as a certain amount of board-feet (a unit for measuring lumber-volume), or as a mechanistic complex of ecological systems that provide previously unmeasured services to the human. In environmental policy, ecosystem services and cost-benefit analysis have been used as metrics to determine how a plant or animal species contributes to human welfare in a quantifiable

 https://doi.org/10.11647/OBP.0186.22

way. These modes of commensuration may provide invaluable bridges into the business community for bringing environmental issues to their attention for serious consideration. Moreover, ecosystem services analysis certainly manifests a form of the transformation of consciousness urgently needed at this time. However, it is also important to ask if such rational perspectives that transform reality into information — namely, manageable, quantifiable data — alter or eliminate other significant ways of knowing reality in relation to decision-making.

One long-term effect is that the individual human decision-maker is distanced from nature because nature is reduced to measurable entities. From this perspective, humans become isolated in our perceived uniqueness as something separate from the biological web of life. In this context, humans do not seek identity and meaning in the numinous beauty of the world, or experience themselves as dependent on a complex of life-supporting interactions of air, water, and soil. Rather, this logic sees humans as independent, rational decision-makers, who find their meaning and identity in systems of management, that now attempt to co-opt the language of conservation and environmental concern. It is a short step within this commensurate worldview to psychological reflection on happiness as personal power derived from simply managing or having more 'stuff'.

This modern, mechanistic, utilitarian view of matter as material for human use and benefit arises in part from a dualistic Western philosophical view of mind and matter. Adapted into Jewish, Christian, and Islamic religious perspectives, this dualism associates mind with the soul as a transcendent spiritual entity given sovereignty and absolute control over wild matter. Many traditional values embedded in religions, such as the sacred, the placement of the sacred in particular geographical locations, the spiritual dimension of the human, and care for future generations, are incommensurate with an objectified reality, and are not quantifiable. Thus, they are often ignored as externalities, or overridden by more pragmatic, profit-driven, bottom line considerations.

Yet, even within the realm of scientific, rational thought, there is not a uniform approach. Resistance to the easy marriage of applied science and instrumental rationality comes from what we might call 'science-that-sees-the-whole.' By this we refer to a lineage embedded in the world of empirical, experimental science of valuing wonder, beauty, elegance, and imagination as crucial components of knowing the world. Knowing, within these perspectives, stresses both analysis and synthesis — the reductive act of observation, as well as placement of the focus of study within the context of a larger whole. 'Science-that-sees-the-whole' resists the temptation to take the micro, empirical, reductive act as the complete description of a thing, but opens analysis to the history of a large interactive web of life. It helps to illustrate the radical interdependence of life that characterizes all ecosystems.

From the Enlightenment period in Western Europe some three centuries ago, the human community has increasingly gravitated towards rational, scientific ways of knowing the world. Modern mechanistic worldviews engender value orientations that separate humans from the Earth. Simultaneously, modernity encourages the primacy of human extractive use and dominion over material reality. The Enlightenment legacy emphasizes knowing the world rationally and scientifically, not religiously or ethically. Rather, religion in modernity orients one away from the immanent and towards the transcendent; whereas ethics examines behavior between humans or between humans and the divine. Moreover, in its economic dimensions, modern worldviews rationalize nature. In this sense, the world at large is without intrinsic value, unless it is calibrated in a metric based on its use value for humans.

This human capacity to imagine and implement a utilitarian-based worldview regarding nature has undermined many insights from the ancient wisdom of the world's religions by segmenting any meaningful religious values as psychological choices or subjective interests. More insidiously, some religions, allured by the individualistic orientations of market rationalism and short-term benefits of social improvement, seized upon wealth and material accumulation as containing divine approval. Thus, early in the nineteenth century, Max Weber identified the rise of Protestant Christianity in Northern Europe with an ethos of inspirited work and accumulated capital. Interestingly, Weber also articulated a disenchantment from the world as this rational, analytical, profit-driven worldview became dominant as global capitalism.

The prior enchantments of the old creation stories were burned away in the critical fires of rationality. Wonder, beauty, and imagination as ways of knowing were gradually superseded in a turn from the organic wisdom of traditional worldviews to the analytical reductionism of modernity. A mercantile mindset sought to shift the play and sport of the world in ways that accorded with modern industrial productivity as the epitome of progress.

Ways of Knowing the World

Certainly, the insights of scientific, analytical, and rational modes of knowing are indispensable for understanding and responding to our contemporary environmental crisis. So also, we will not bring ourselves out of our current impasse without the technologies that brought us into it. Indeed, these technologies are being reshaped in more ecological directions as witnessed in such developments as industrial ecology and green chemistry. But it seems important also to recall that other ways of knowing are manifest in culturally diverse cognitive pathways that treasure emotional intelligence and affective insight. These are evident in the arts — music, painting, literature, poetry, drama — that celebrate human experience in a more than rational mode. Moreover, in their explorations of embodied experience of humans and nature, many aspects of Western culture, such as visual aesthetics, literary arts, narrative poetry, and cinema are far from dormant in modern consciousness.

What is especially striking in this regard are the versions of empirical observation found among Indigenous, or aboriginal, peoples that have both rational and affective components. This involves knowledge of lands and ocean, animals and fish, plants and trees. These many ways of knowing appear in an amazing variety of human interactions with the natural world that include: the development of traditional herbal knowledge, proto-chemical understandings, healing practices, philosophical reflection in oral-narrative traditions, and agricultural cultivation. These diverse ways of knowing-dialogues are evident in the domestication of various crops such as rice, millet, wheat, corn, and tobacco. Much of modern science was built upon these foundational insights. Such understandings must have come through a wide range of careful observation and attention to seasonal changes and animal interactions. Similar observational knowledge of the migratory patterns of plants, animals, birds, and fish is evident among many native cultures. Almost uniformly, the remaining Indigenous oral narratives describe this trial and error in experimental usage along with inspired reflection on the beauty and profundity of an in-spirited world. One insight is that many modes of Indigenous knowledge often refer to these connections with the world as kin relationships.

Thus, it is becoming clearer that new modes of integrating traditional environmental knowledge and science are emerging.

> Bridging multiple knowledge systems requires drawing on natural and social sciences' methodologies and constant consideration for the value systems of all knowledge holders, a process that is based on ongoing iteration and feedback. The Mi'kmaq principle of 'Etuaptmumk' or 'two-eyed seeing' captures the concept of bringing different knowledge systems together to increase our collective bread and depth of understanding: 'learning to see from one eye with the strengths of indigenous knowledges… and from the other eye with the strengths of western knowledges…and learning to use both these eyes together, for the benefit of all'.[1]

Science-that-sees-the-whole is beginning to appreciate these other ways of knowing without giving over its foundational analytical approach. In recent years, science has returned to study Indigenous knowledge not simply as idiosyncratic experiences, but as connected to larger social and ecological phenomena. Increasingly, these connections are understood as creative entanglements of the senses and the cognitive faculties. Over the last century, new ways of understanding reality have moved from the periphery of our knowledge into more common usage that increasingly tip us toward creative engagement with cosmology. For example, ways of seeing reality at the quantum level as simultaneously particle and wave, as multicentered, and as foaming into and out of existence are beginning to challenge creatively our articulation of everyday life. Our mental horizon now embraces the comprehensible and the intuitive in ways that formerly would have been dismissed as contradictory or logically incompatible.

In addition, there is a growing appreciation for the multiple 'intelligences' in the world. This book aims to explore some of those

---

1   Susan Kutz and Matilde Tomaselli, 'Two-Eyed Seeing Supports Wildlife Health', *Science*, 364.6446 (2019), 1135–37, at 1136, https://doi.org/10.1126/science.aau6170; inner quote from Shelley K. Denny, Lucia M. Fanning, 'A Mi'kmaw Perspective on Advancing Salmon Governance in Nova Scotia, Canada: Setting the Stage for Collaborative Co-Existence', *International Indigenous Policy Journal*, 7.3 (2016), 1–25, at 16, https://doi.org/10.18584/iipj.2016.7.3.4

intelligences from plants and animals, to trees and forests. It recognizes both Indigenous ways of knowing and modern ecological ways of knowing. In both cases, organic interconnectivity is acknowledged and affirmed. Those participating in this book bring an appreciation for multiple ways of knowing and multiple intelligences in the world. Their work reflects the careful attempt to 'see the whole'. Our work collaboratively aims to bring that sensibility toward an embodied ethic for nature.

Bibliography

Abram, David, *The Spell of the Sensuous: Perception and Language in a More-Than-Human World* (New York, NY: Pantheon Books, 1996).

Bachelard, Gaston, *Water and Dreams: An Essay on the Imagination of Matter*, trans. by Edith R. Farrell (Dallas, TX: The Dallas Institute of Humanities and Culture, 1983).

Basso, Keith, *Wisdom Sits in Places: Landscape and Language among the Western Apache* (Albuquerque, NM: University of New Mexico Press, 1996).

Berkes, Fikret, *Sacred Ecology*, 2nd ed., (New York, NY, and London: Routledge, 2008), https://doi.org/10.4324/9780203928950

Berry, Thomas, *The Dream of the Earth* (San Francisco and Berkeley, CA: Sierra Club Books, 1988).

— *The Great Work* (New York, NY: Bell Tower, 1992).

— *Sacred Universe: Earth Spirituality and Religion in the 21st Century*, ed. by Mary Evelyn Tucker (New York, NY: Columbia University Press, 2009).

Bourdieu, Pierre, *The Logic of Practice*, trans. by Richard Nice (Stanford, CA: Stanford University Press, 1990).

Cordova, V. F., *How It Is: The Native American Philosophy of V. F. Cordova*, ed. by Kathleen Dean Moore, Kurt Peters, Ted Jojola, and Amber Lacy (Tucson, AZ: The University of Arizona Press, 2007).

Damasio, Antonio, *Descartes' Error: Emotion, Reason, and the Human Brain* (New York, NY: Grosset/Putnam, 1994).

— *Looking for Spinoza: Joy, Sorrow, and the Feeling Brain* (New York, NY: Harcourt, 2003).

Denny, Shelly K., and Lucia M. Fanning, 'A Mi'kmaw Perspective on Advancing Salmon Governance in Nova Scotia, Canada: Setting the Stage for Collaborative Co-Existence', *International Indigenous Policy Journal*, 7.3 (2016), 1–25, https://doi.org/10.18584/iipj.2016.7.3.4

Dewey, John, *Experience and Nature* (New York, NY: W. W. Norton and Co., [orig. 1925 edition, Paul Carus Lecture] 1929).

Glacken, Clarence J., *Traces on the Rhodian Shore: Nature and Culture in Western Thought from Ancient Times to the End of the Eighteenth Century* (Berkeley, CA: University of California Press, 1967).

Goodenough, Ursula, *The Sacred Depths of Nature* (New York, NY: Oxford University Press, 1998).

Grim, John, and Mary Evelyn Tucker, *Ecology and Religion* (Washington, DC: Island Press, 2014).

Grim, John., ed., *Indigenous Traditions and Ecology: The Interbeing of Cosmology and Community* (Cambridge, MA: Harvard Divinity School Center for the Study of World Religions, 2001).

Grim, John, *The Shaman: Patterns of Religious Healing Among the Ojibway Indians* (Norman, OK: University of Oklahoma Press, 1983).

Harrod, Howard, *The Animals Came Dancing: Native American Sacred Ecology and Animal Kinship* (Tucson, AZ: The University of Arizona Press, 2000).

Heidegger, Martin, *Poetry, Language, Thought*, trans. by Albert Hofstadter (New York, NY: Harper & Row, 1971).

Holmes, Barbara, *Race and the Cosmos* (Harrisburg, PA: Trinity Press International, 2002).

Ingold, Tim, *The Perception of the Environment: Essays in Livelihood, Dwelling and Skill* (London: Routledge, 2000).

Jenkins, Willis, Mary Evelyn Tucker, and John Grim, eds, *Routledge Handbook of Religion and Ecology* (New York, NY: Routledge, 2016), https://doi.org/10.4324/9781315764788

Johnson, Mark, *The Body in the Mind: The Bodily Basis of Meaning, Imagination, and Reason* (Chicago, IL, and London: University of Chicago Press, 1987).

Jullien, François, *Vital Nourishment: Departing from Happiness*, trans. by Arthur Goldhammer (New York, NY: Zone Books, 2007).

Kellert, Stephen, and Ed Wilson, eds, *The Biophilia Hypothesis* (Washington, DC: Island Press, 1993).

Kellert, Stephen, and Timothy Farnham, eds, *The Good in Nature and Humanity: Connecting Science, Religion, and Spirituality with the Natural World* (Washington, DC: Island Press, 2002).

Kutz, Susan, and Matilde Tomaselli, 'Two-Eyed Seeing Supports Wildlife Health', *Science*, 364.6446 (2019), 1135–37, https://doi.org/10.1126/science.aau6170

Lakoff, George, and Mark Johnson, *Philosophy in the Flesh: The Embodied Mind and Its Challenge to Western Thought* (New York, NY: Basic Books, 1999).

Levinas, Emmanuel, *Otherwise than Being or Beyond Essence*, trans. by Alphonso Lingis (Boston, MA: M. Nijhoff, 1981).

Marion, Jean-Luc, *Being Given: Toward a Phenomenology of Givenness*, trans. by Jeffrey Kosky (Stanford, CA: Stanford University Press, 2002).

Merleau-Ponty, Maurice, *Phenomenology of Perception* (London: Routledge and Egan, 1962).

Midgley, Mary, *The Myths We Live By* (London and New York, NY: Routledge, 2003).

Milton, Kay, *Loving Nature: Towards an Ecology of Emotion* (London and New York, NY: Routledge, 2002).

Nabhan, Gary Paul, and Stephen Trimble, The Geography of Childhood (Boston, MA: Beacon Press, 1994).

Nancy, Jean-Luc, *The Ground of the Image*, trans. by Jeff Fort (New York, NY: Fordham University Press, 2005).

Slingerland, Edward, 'Embodying Culture: Grounding Cultural Variation in the Body', in *What Science Offers the Humanities* (Cambridge, UK: Cambridge University Press, 2008), pp. 151–218, https://doi.org/10.1017/cbo9780511841163.007

Swimme, Brian, *The Hidden Heart of the Cosmos: Humanity and the New Story*, rev. ed. (Maryknoll, NY: Orbis Books, 2019).

Swimme, Brian, and Thomas Berry, *The Universe Story* (San Francisco, CA: Harper Collins, 1992).

Swimme, Brian, and Mary Evelyn Tucker, *Journey of the Universe* (New Haven, CT: Yale University Press, 2011).

Teilhard de Chardin, Pierre, *The Human Phenomenon*, trans. by Sarah Appletone-Weber (Brighton, UK: Sussex Academic Press, 1999).

Toulmin, Stephen, *The Return to Cosmology: Postmodern Science and the Theology of Nature* (Berkeley, CA: University of California Press, 1982).

Tu Weiming and Mary Evelyn Tucker, eds, *Confucian Spirituality*, 2 vols (New York, NY: Crossroads, 2003, 2004).

Tucker, Mary Evelyn, *Worldly Wonder: Religions Enter their Ecological Phase* (Chicago, IL: Open Court, 2002).

Yellowtail, Thomas, *Yellowtail: Crow Medicine Man and Sundance Chief, as Told to Michael O. Fitzgerald* (Norman, OK: University of Oklahoma Press, 1991).

Zizioulas, John (of Pergamon), *Remembering the Future: An Eschatological Ontology* (London: T&T Clark, 2009).

SECTION I

PRESENCES IN THE
MORE-THAN-HUMAN WORLD

Fig. 2 Sandhill Crane in Flight, near Muleshoe National Wildlife Refuge, Baily Country, Texas. Photo by Leaflet (2009), Wikimedia, CC BY-SA 3.0, https://commons.wikimedia.org/wiki/Grus_canadensis#/media/File:Sandhill_Crane_Bailey_County_Texas_2009.jpg

1. Creaturely Migrations on a Breathing Planet

Some Reflections[1]

David Abram

Not so long ago, almost every stream, river, and rill emptying into the North Pacific — from southern Japan to northern Siberia, around the huge Alaskan land mass and the Pacific coast of Canada all the way to southern California — supported one or more salmon runs, perfectly adapted to the dynamic ecology of that watershed. The migratory ways of the salmon were likely born in response to the swelling and subsiding ice ages that have dominated their range after these fish evolved to their present form in the cold freshwater of the northern latitudes some two million years ago. As those inland waters were subsumed beneath the immense, spreading ice sheets, the fish were driven out into the ocean and forced to adapt, yet, somehow, they never lost their ancestral tie to the fresh mountain streams.

Whenever the ice sheets retreated, salmon would colonize the rivers and tributaries formed by the glacial runoff, slowly establishing new spawning beds in valleys scoured out and scraped by the ice. They brought with them the rich nutrients of the open ocean, and, as bear, eagle, and otter feasted upon their spawning or spawned-out bodies,

---

1   This chapter is an excerpt from 'Creaturely Migrations on a Breathing Planet' — a long essay reflecting upon the wild Sandhill crane, Pacific salmon, and Monarch butterfly migrations, published in the inaugural issue of *Emergence* (an online magazine launched on Earth Day, 2018). The full essay is published there in both textual and auditory form (read by the author): https://emergencemagazine.org/story/creaturely-migrations-breathing-planet

          https://doi.org/10.11647/OBP.0186.01

this abundant ocean nourishment was distributed more widely, enriching the soils and enabling the sparse, pioneer woodlands to fill out into dense forests.

These forests, in turn, shaded the inland streams, their detritus providing shelter and food for the aquatic insects and small fish upon whom the salmon themselves fed. The thickening woods offered habitat for innumerable other animals, for raven and coyote, for owl and deer and raccoon. The association between the salmon and woodlands is an ancient reciprocity renewed time and again: as the ice retreated, the fish and the forests recolonized the glacier-scoured terrain together.

After the most recent ice age, nomadic bands of humans, as well, made their way up into the coastal forests and the river valleys, drawing sustenance from the seasonal storms of large fish that would periodically undulate up the rivers, surging past fallen trunks and leaping up waterfalls. Centuries later, while developing ways of preserving the caught salmon (drying them in the wind or smoking them on wooden racks), some native peoples settled in permanent villages along those rivers. Through trial and calamitous error — at times overfishing the runs and having to endure the consequent seasons of famine — these cultures gradually learned how best to harvest the collective gift without interrupting its cyclical replenishment. Central to *all* such cultural constraints, throughout the Pacific Rim, was a recognition of the salmon as a powerful emissary from hidden or unseen dimensions — a form of energetic intelligence that came toward humankind from the sacred heart of the mysterious.

Indigenous cultures from every part of the North Pacific Rim revered the salmon as an uncommonly holy power, ritualizing their respect in ceremonies that honored the first salmon caught in the spring. No other salmon could be taken during these rites, wherein the first salmon was treated as an esteemed guest before being carefully prepared and eaten.

Along the Klamath River in what is now California, the primary 'first salmon ceremony' was conducted in a Yurok village close by the mouth of the river. After that event, strong runners were sent upriver to alert the Hoopa people that the proper rites had been accomplished, and that the spring salmon were on their way. Upon catching their own first fish, the Hoopa undertook ten days of ceremony and prayer before allowing generalized fishing to commence. The Karuk people, many miles further upstream, moved away from the river and into the hills while the first salmon was taken and ritually eaten by their spiritual leaders.

One effect of such ceremonies, and of the restrictions on fishing during their enactment, was that significant numbers of early salmon were enabled to pass freely upriver to their spawning grounds, ensuring the continued replenishment of the run. The ritualized honoring of the first fish also ensured that the salmon, however abundant in the coming season, could not be taken for granted — that its flesh remained a sacrament for the people.

Across the Pacific, among the Ainu people of northern Japan, whenever any family caught the season's first salmon from the river, the fish was passed through a special window into the house before being placed in front of the hearth fire. There the family would address the spirit of the salmon directly, honoring it ceremonially with spoken words and ritual gestures. The household fire, for the Ainu, was itself a goddess who could see all that unfolded around her; she would report back to the other gods that the salmon had been treated with proper respect.

The Ainu held ceremonies, too, to bid goodbye to the salmon when, having left their flesh bodies behind as food, they paddled their spirit-boats back to their homes far to the east. Like other native peoples of the North Pacific, the Ainu assumed that the salmon, when they were not crowding upstream to visit the people, removed their salmon garb and lived in human form beyond the ocean horizon. Such a mythic view bound their sensory imagination to the ways of this wild creature, engendering an almost familial regard for its well-being. The view was so widespread that, when in the nineteenth century when several Skagit Indians from the American northwest accompanied a white expedition back to the east coast, and saw the abundance of pale, pink-skinned people living there, they reported back to their tribe that they had been to salmon country and had seen the salmon walking around as human beings.[2]

More detached and technological approaches to tracking salmon have yielded other ways of describing their whereabouts once they depart

---

2   June McCormick Collins, 'The Mythological Basis for Attitudes towards Animals among Salish-Speaking Indians', *Journal of American Folklore*, 65 (1952), 353–59, https://doi.org/10.2307/536039

14 *Living Earth Community*

the inland waters. Upon leaving their rivers, the salmon seem to spend the largest part of their lives swimming in great circles throughout the North Pacific. Their journeys carry them to the remotest regions of the sea, feeding and growing strong on the ocean's abundance — on herring and smelt and other small fish — traveling distances that boggle the human mind. After several years dispersing to all points on the horizon, following their food whence it leads them, the members of a single run unerringly return to the mouth of their natal stream — all converging there, somehow, at precisely the same time. How they pull off this feat remains an enigma for present-day science. Once the salmon come close to their home stream, it is probable that they rely on their astonishingly keen sense of smell to distinguish between the subtly different waters of neighboring tributaries. But how the fish navigate across thousands of miles of ever-shifting and largely featureless ocean to make their way back to the very same coastal point from whence they set out years earlier, remains an elemental mystery to us, confounding our primate senses and our terrestrial, pedestrian logic.

Like spring monarch butterflies fluttering north toward specific clumps of milkweed that only their great-grandchildren will reach, like sandhill cranes vibrating the sky with their bugling as they drop toward a tiny patch of peat bog in the broad tundra, the migrating salmon appear to avail themselves of somatic skills far beyond our bodily ken. The only way contemporary science seems able to fathom their uncanny navigational powers is by likening the abilities of these animals to technologies of our own, human invention. We are told, over and again, that these migratory creatures make use of internal maps and internal compasses, of innate calendars and inborn clocks. Clocks, compasses, and calendars, however, are by definition *external* contrivances, ingeniously built tools that we deploy at will. Metaphorically attributing such instrumentation to other animals has confounding implications, suggesting a curious *doubleness* in the other creature — a separated sentience or self that regularly steps back, within its body or brain, to consult the map or the calendar.

It seems unlikely, however, that organisms interact with an internal representation of the land in any manner resembling our own engagement with maps. Cranes and butterflies would have little use for a separated *re-presentation* of the earth's surface, for they have never torn themselves out of the encompassing *presence* of the wide earth. Our reliance upon such instrumental metaphors seems to stem from our over-civilized assumption of a neat distinction between living organisms and the non-living terrain that they inhabit, an unambiguous divide between animate life and the ostensibly inanimate planet on which life happens to locate itself. As long as the material ground is considered inert — as long as the elemental atmosphere or ocean is viewed as a passive substrate — then the long-range migrations of certain animals can only be a conundrum, a puzzle we will strive to solve by continually compounding various internal mechanisms that might somehow, in combination, grant a particular creature the power to grapple its way across the world. Instead of hypothesizing more metaphorical gadgets, adding further accessories to a crane's or a salmon's interior array of tools, what if we were to allow that the animal's migratory skill arises from a felt rapport between its body and the breathing Earth? That a crane's two thousand mile journey across the span of a continent is propelled by the felt unison between its flexing muscles and the sensitive Flesh of this planet (this huge curved expanse, roiling with air currents and rippling with electromagnetic pulses), and so is enacted as much by Earth's vitality as by the bird that flies within it?

Such a conception need not contradict any of the accepted evidence gathered from a century's research into the enigmas of animal migration; it simply offers a new way of interpreting and integrating those various evidences. By focusing our questions so intently on the organism, as if it carries all the secrets of this magic hidden inside itself, we easily lose sight of the obvious collaboration at play. By adding new gadgets to an animal's neurological and genetic endowment, we tacitly induce ourselves to focus upon relationships *interior* to the organism (how, for example, does the animal bring its biological clock and its internal map to bear on its compass readings), deflecting our curiosity and attention from the more mysterious relationship that calls such interactions into being.

What is this dynamic alliance between an animal and the animate orb that gives it breath? What seasonal tensions and relaxations in the atmosphere, what subtle torsions in the geosphere help to draw half a million cranes so precisely across the continent? What rolling sequence or succession of blossomings helps summon these millions of butterflies across the belly of the land? What alterations in the olfactory medium, what bursts of solar exuberance through the magnetosphere, what attractions and repulsions...? For surely, really and truly, these migratory creatures are not taking readings from technical instruments nor mathematically calculating angles; they are riding waves of sensation, responding attentively to allurements and gestures in the topological manifold, reverberating subtle expressions that reach them from afar. These beings are dancing not with themselves but with the animate rondure of the earth, their wider Flesh!

Consider the deep somatic attunement by which a salmon feels its way between faint electromagnetic anomalies, riding a particular angle of sun as it filters down through the rippled surface, gliding with certain currents and plunging up against others, dreaming its way through gradients of scent and taste toward a particular bend of gravel and streamside shadow. Whatever specialized sensitivities and internal organs are brought to bear, those very organs have co-evolved with textured patterns and pulses actively propagating through the elemental medium; indeed, those sensitivities have often been *provoked* by large-scale repetitive or rhythmic happenings proper to that part of the biosphere — by pulsed coalescences and cyclic dispersals — and so can hardly be fathomed without reference to these patterned gestures within the Body of the planet.

Perhaps it would be useful, now and then, to consider the large, collective migrations of various creatures as active *expressions* of the earth itself. To consider them as slow gestures of a living geology, improvisational experiments that gradually stabilized into habits now necessary to the ongoing metabolism of the sphere. For truly: are not these cyclical pilgrimages — these huge, creaturely hegiras — also pulsations within the broad Body of earth? Are they not ways that divergent places or ecosystems communicate with one another, trading vital qualities essential to their continued flourishing?

Think again of the salmon, this gift born of the rocky gravels and melting glaciers, nurtured by colossal cedars and by tumbled trunks decked with ferns, fungi, and moss, an aquatic, muscled energy strengthening itself in the mossed and forested mountains until it's ready to be released into the broad ocean. Pouring seaward, it adds itself to that voluminous cauldron of currents spiraling in huge gyres, shaded by algal blooms and charged by faint glissandos of whale song... Until, grown large with the sea's abundance, this ocean-infused life flows back up the rivers and tributaries and spreads out into the wooded valleys, gifting the hollows and the needled highlands with new minerals and nutrients, feeding bears and osprey and eagles, ensuring that the glinting gift will be reborn afresh from a lump of luminous eggs stashed beneath a layer of pebbles.

This circulation, this *systole and diastole*,[3] is one of the surest signs that this *earth is alive* — a rhythmic pulse of silvery, glacier-fed brilliance pouring through various arteries into the wide body of the ocean, circulating and growing there, only to return by various veins to the beating heart of the forest, gravid with new life.

Or... perhaps it's better to think of this seasonal reciprocity as a kind of *breathing*, as an exhalation of millions of young salmon smolts down from the tree-thick mountains and meadows and then out into the roiling cosmos of currents and tidal flows, to mingle with zooplankton and seals and squids, and then the great in-breath, the drawing in of living nourishment from the sea into river mouths and estuaries, inhaling the salmon up those rivers into streams, and from there into the branching becks, rills, and runnels that filter into the green forests, the living lungs of this biosphere. Or is it the broad-bellied ocean that is breathing, sucking these finned nutrients down from the shaded slopes, luring them over rocks and through rapids and hydroelectric dam spillways — drawing them past bustling cities and factories, through intersecting gradients of toxic effluents that sting their mouths and strafe their exposed membranes — on out into the heaving whirl of the sea's innards, the writhe and foam of the ocean circulating this glimmering nourishment within itself before exhaling it back, a long sighing breath, up into the wooded valleys?

However else we may view them, these deterritorializations and reterritorializations — these large migrations of various species — are a primary way that the biosphere cleanses and flexes its various organs, replenishing itself: each region drawing insights from the others, concentrating and transforming such qualities before releasing them abroad, divergent places trading perspectives along with nutrients and

---

3   The two phases of the cardiac cycle.

nucleotides, the whole half-shadowed sphere steadily experimenting, improvising, slowly altering its display to the blazing fire watching from afar, as the reflective moon rolls on 'round.

Bibliography

Collins, June McCormick, 'The Mythological Basis for Attitudes towards Animals among Salish-Speaking Indians', *Journal of American Folklore*, 65 (1952), 353–59, https://doi.org/10.2307/536039

2. Learning a Dead Birdsong

Hopes' echoEscape.1 in 'The Place Where You Go to Listen'

Julianne Lutz Warren

Prelude

It was spring of 2011. I was searching for something else in Cornell University's archive of sounds when I first came across a sixty-three-year-old recording of a Māori man whistling his memory of songs of Huia.[1] These tones both cut and enchanted me. In 1948, these birds were already believed to be extinct. Huia — whose distress notes speak in their onomatopoeic given name[2] — were endemic to Aotearoa New Zealand. The elder Huia mimic — Henare Hamana (aka Henare 'Harry' Salmon 1880–1973) of the Ngati Awa hapu of Warahoe — had been invited into a Wellington recording studio by a Pākehā,[3] a neighbor called R. A. L. (Tony) Batley (1923–2004) (see Figure 3.1). Batley, who also narrated the recording, was a regional historian from Moawhango settlement. He was interested in preserving this remnant of remembered avian language. The birds were, as Batley puts it in the recording, 'of unusual interest'. For instance, all Huia had ivory-colored bills, but they were curiously dimorphic in shape. The bills belonging to males were generally shorter and more like 'pick-axes' than females' bills, which

---

1   R. A. L. Batley, Archives, Box 2011.117.1 MS 177, Whanganui Regional Museum. The original recording can be heard here: https://macaulaylibrary.org/asset/16209

2   Michael Szabo, 'Huia, The Sacred Bird', *New Zealand Geographic*, 20 October-December 1993, https://www.nzgeo.com/stories/huia-the-sacred-bird

3   The Māori-language term for a white person, typically of European descent.

      https://doi.org/10.11647/OBP.0186.02

were long and curving. Both sexes were crow-size, with black-green bodies and a dozen stiff tail feathers edged, again, in ivory. Against the darkness of dense native trees and ferns filtering sunlight, the bright trim on each bird leaping between low limb and earth must have arced like coupled meteorites through a night sky.

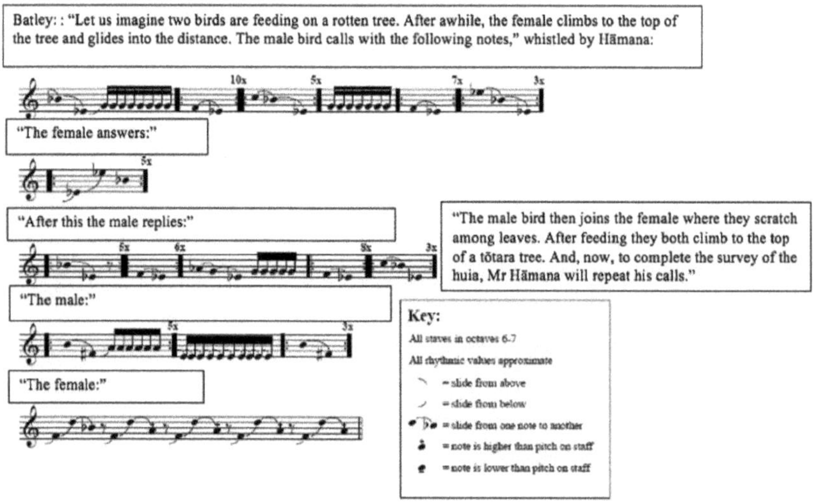

Fig. 3.1 Transcription of R. A. L. (Tony) Batley's recording, by Dr. Martin Fellows Hatch, Emeritus Professor, Musicology, and Dr. Christopher J. Miller, Senior Lecturer/Performer, Cornell University (2015).

Huia became extinct due to complex human causes that were local, but globally common due to expanding colonization. Huia range likely contracted, then stabilized, after the arrival of Māori ancestors around a millennium ago. A couple of hundred years of European settlement escalated stresses on the birds. Ecological communities of old-growth forests co-evolved with Huia were widely cut-down and disrupted by newcomers' with industrial and capitalist assumptions. These underpinned an overpowering system of cropping and livestock-grazing alongside an influx of unfamiliar avian predators and parasites.[4] These

---

4   For discussions of extinctions in the wake of both Māori and Pākehā arrival, see, for example: Alan Tennyson and Paul Martinson, *Extinct Birds of New Zealand* (Wellington: Te Papa Press, 2007) and Atholl Anderson, 'A Fragile Plenty: Pre-European Māori and the New Zealand Environment', in *Environmental Histories of New Zealand* , ed. by Eric Pawson and Tom Brooking (Oxford: Oxford University Press, 2002), pp. 35–51; see also Catherine Knight, *Ravaged Beauty: An Environmental History of the Manawatu* (Ashhhurst: Totara Press, 2014).

wide-scale harms in turn alienated Māori from customary relationships as Tangata Whenua[5] (People of the Land), a worldview basis of cultural identity and Indigenous authority entwined with language and the health of the land, Huia included. Generations of Māori had learned to attract Huia, who were tapu[6] or sacred, by imitating the birds' voices. Māori had ritually snared Huia for tail feathers — which were sometimes given as gifts — and other ceremonial or ornamental parts. Huia may have sometimes been eaten. These birds emerged in ancient cosmology and conveyed messages in living dreams. With the privileging of British economic valuations, mounted Huia skins and tail feathers became commodified and sold internationally, which increased the rate at which they were killed by both Māori and Pākehā hunters. By the end of the nineteenth century it had become evident to observers across cultures that Huia had been brought to the brink of extinction. Some Māori, particularly Ngāti Huia, placed their own protections on Huia range, protecting the birds. In response to the hapu's appeal, in 1892, the Crown government also extended its Wild Bird Protection Act to Huia.[7] One idea set forth by this government officials was to capture Huia pairs in their North Island home range and move them to offshore island sanctuaries: Hauturu Little Barrier Island and Kapiti Island were possible destinations.[8]

---

5   In te reo Māori — English translations of 'whenua' include, land, country, ground, and placenta. Margaret E. Forster, 'Recognising Indigenous Environmental Interests: Lessons form Aotearoa New Zealand,' Presented at *International Political Science Association, Research Committee 14: Politics and Ethnicity. The Politics of Indigenous Identity: National and International Perspectives*. Dunmore Lang College, Macquarie University, Sydney (July, 2013). Rachel Selby, Malcolm Mulholland, and Pataka Moore, eds, *Maori and the Environment* (Wellington: Huia Publishers, 2010); Trudie Cain, Ella Kahu, and Richard Shaw, eds, *Turangawaewae: Identity and Belonging in Aotearoa New Zealand* (Auckland: Massey University Press, 2017).

6   English translations of 'tapu' include 'set apart' or 'under restriction'.

7   Szabo, 'Huia, The Sacred Bird.'

8   For discussion of huia-human interrelations, see: W. J. Phillips, *The Book of the Huia* (Christchurch: Whitcombe and Tombs Ltd, 1963); W. J. Phillips, 'Huia Research: 1927–1954, (MU000235)', Dominion Museum (creating agency) (Wellington: Museum of New Zealand/Te Papa Tongarewa, [n.d.]); H. T. Whatahoro, *The Lore of the Whare-Wānaga: Or Teachings of the Māori College on Religion, Cosmogony, and History Vol. 1: Te Kauwae-Runga, or 'Things Celestial'*, trans. by S. Percy Smith (Cambridge, UK: Cambridge University Press, 2011), https://doi.org/10.1017/CBO9781139109277; Margaret Orbell, *The Natural World of the Māori* (Auckland: Collins, 1985); Murdoch Riley, *Māori Bird Lore* (Paraparaumu, NZ: Viking Sevenseas, NZ Ltd, 2001); Walter L. Buller, *A History of the Birds of New Zealand, Vol. 1* (London: self-published, 1888), pp. 9–17; Geoff Norman, *Bird Stories: A History of the Birds of New Zealand* (Nelson: Potton and Burton, 2018).

Fig. 3.2 *NGA HURUHURU RANGATIRA*, sculpture by Robert Jahnke (2016), Palmerston North, New Zealand. Photograph by Julianne Lutz Warren (2016).

By the time expeditions were arranged with this intent of rescuing Huia, it was apparently too late. Both Batley and Hamana had been engaged in one or more Huia searches led by Dominion Museum officials (now Te Papa Tongarewa). Hamana was known for his bush skills, including imitating the birds. In 1909, he joined one of those Crown-official parties tramping into the still-dense forest in the Mt. Aorangi-Mangatera Stream area some miles from Taihape.[9] Reports of this expedition vary. Someone might have seen or heard Huia call on this outing. No living Huia pair was obtained by Museum staff, however, during any of the trips. Claims of encountering Huia lingering here and in other hard-to-reach strongholds throughout the North Island continued even past mid-century, several convincing.[10]

---

9   Under the Crown government most of this forested area was in blocks collectively-owned by Māori, in particular, by members of Hamana's Iwi.

10   Phillips, *The Book of the Huia* and 'Huia Research'; Dean Andrew Baigent-Mercer, 'Brief of Evidence,' In the Waitangi Tribunal Wai 1040 Te Paparahi o Te Raki Inquiry District Wai 1661, 31 October 2016, 25.

Around the same time that I first heard the bird-man recording, I also encountered 'Learning a Dead Language', a poem of five stanzas by poet W. S. Merwin.[11] Via the ambiguity it conveyed I felt released from too-simple, clear-cut norms eroding so many hopes of flourishing lives. This poem ushered me into a tangled bank fertile with alternative narratives of trust and desire.

In its first stanza, 'Learning a Dead Language' suggests,

> There is nothing for you to say. You must
> Learn first to listen. Because it is dead
> It will not come to you of itself, nor would you
> Of yourself master it. You must therefore
> Learn to be still when it is imparted,
> And, though you may not yet understand, to remember.

At the beginning, I imagine now, Ancestors meeting Unborn. Learning stillness, the listener feels how cold she is somewhere in-between.

> What you remember is saved. To understand
> The least thing fully you would have to perceive
> The whole grammar in all its accidence
> And all its system, in the perfect singleness
> Of intention it has because it is dead.
> You can learn only a part at a time.

What may come of this double-bind in which a listener 'can learn only a part at a time,' yet, disconnected from its 'whole grammar,' 'The least thing' cannot be 'understood... fully'?'

Perhaps, I sense — as humility postures — understanding, like becoming a forest, is group work.

> What you are given to remember
> Has been saved before you from death's dullness by
> Remembering. The unique intention
> Of a language whose speech has died is order,
> Incomplete only where someone has forgotten.
> You will find that order helps you to remember.

---

11 W. S. Merwin, 'Learning a Dead Language', in *Migration: New and Selected Poems* (Port Townsend, WA: Copper Canyon Press, 2005), p. 41. See also, Julianne Warren, 'Hopes Echo', The Poetry Lab, *The Merwin Conservancy*, 2 November 2015, https://merwinconservancy.org/2015/11/the-poetry-lab-hopes-echo-by-author-julianne-warren-center-for-humans-and-nature

Who else is remembering? I can hear generations of learners — stolen children — convened and puzzling out their relations as kin. Each listener contributes saved pieces of a torn up composition, filling in gaps for one another. There are voids that remain.

> What you remember becomes yourself.
> Learning will be to cultivate the awareness
> Of that governing order, now pure of the passions
> It composed; till, seeking it in itself,
> You may find at last the passion that composed it,
> Hear it both in its speech and in yourself.

From whispered mutters each learner hears their own voice welcoming and welcomed by each other's, communication re/emerges, an indisputable chorus.

> What you remember saves you. To remember
> Is not to rehearse, but to hear what never
> Has fallen silent. So your learning is,
> From the dead, order, and what sense of yourself
> Is memorable, what passion may be heard
> When there is nothing for you to say.

At the end, as it was in the beginning, '…there is nothing for you to say.' Yet, the listener is warmer now — while still — in another round of learning, and, another, of deepening cycles of communion with fluent pasts and futures joining into an ever-speechless, but not as lonely, present.

It seemed to me that this poet limned my sense of strengthening, yet obscure loneliness at the same time brightening dimmed expectations for companionship. I was numb with knowledge of the very many kinds of life — and languages — that had recently been rapidly extinguished, as well as the scientific projections of further destruction of life. This numbness bothered me. On the one hand, a lack of feeling made it easier to pretend otherwise. And, on the other hand, such detachment made it easier to weaponize suppressed anger. I wondered whether this difficulty with engagement, which I sensed was widespread, might not be leading otherwise compassionate people not only to abandon hope, but to slander it.[12] This possibility scared me. Part of the problem, I guessed, was the abstract nature of 'extinction' and of

---

12   For example, Derrick Jensen deems hope a 'bane' in 'Beyond Hope', *Orion Magazine*, 2 May 2006, https://orionmagazine.org/article/beyond-hope

'kind' in contrast with the loss of a particular known face or voice. Another part of the problem was self-preservation — who could really bear such a massive intensity of death? For some of us, awakening to challenge harmful assumptions of our own white supremacy, this intensity would be compounded by our complicity. Another part of the problem, I believed, was my dominating culture's systemic alienation (also taking forms of forced assimilation) of other humankinds and of other-than-human community members and divisions of past from future and emotion from reason. These forms of violence must certainly translate into mortal wounds in truth-telling. And, without that, how can anybody express love skillfully?

So, these encounters seemed like something to pay attention to — this poem, this dead bird language, parts of which had been remembered by others who 'saved before you', in Merwin's words. Those others, many Māori, themselves had Te Reo, the language of the Tangata Whenua, 'bashed out of'[13] their own or their parents' mouths in English-only Crown schools. Indeed, at an even higher at-risk proportion than Earth's birdkinds, most languages of humankinds, a majority of Indigenous ones, are at risk of dying with their few remaining speakers.[14] How then, might I properly respond to these saved song remnants as a far-away learning listener? The multiplex voice of the recording, marking absence, haunted my life, becoming, through intimacy, a relied-on presence.[15] I kept revisiting the old recording and re-reading this poem. Eventually, I edited out the English-language narration, recognizing that the imprint of Batley's speech would nonetheless shape Hamana's whistled phrases. Now, however, I could play this compressed conversation between male and female extinct birds recalled by human ancestors — of colonizer

---

13   E. E. Campbell, *Central District Times*, Letter to the Editor, 3 December 1974 in RAL Batley, Archives Box 16, 2011.117.1, Whanganui Museum, NZ.

14   According to BirdLife International, 40% of Earth's birds are in decline. See: https:// e360.yale.edu/digest/forty-percent-of-the-worlds-bird-populations-are-in-decline-new-study-finds and UNESCO predicts, in current circumstances, that 50%–90% of Indigenous languages (approximately 3000) will be replaced by English, Madarin or Spanish by 2100. See https://www.iwgia.org/en/focus/international-year-of-indigenous-languages.html

15   Alan Licht, 'Sound Art: Origins, Development and Ambiguities', *Organised Sound*, 14.1 (2009), 3–10, https://doi.org/10.1017/s1355771809000028. I use the word 'companion' in the same sense as Licht (7). Also, I acknowledge the influence of David Abram, *Becoming Animal: An Earthly Cosmology* (New York, NY: Vintage, 2011), whose work has informed my understanding of relationality.

and colonized — disclosed through machines repeating in a loop. And, I would be less distracted by the recall of my own dominating lexicon. I started hearing this edited sound-cycle as 'Huia Echoes'.

Under the influence also of R. Murray Schafer's *The Soundscape*, I then wondered whether other parts of 'dead' bird and human and other languages might be imparted if I was 'still' while listening to this circling of songs interact within different milieus.[16] The strangeness of this thought captivated me. So, I carried Huia Echoes around in my handheld digital device. I also carried a small, inexpensive recorder, which I used to re-store Huia Echoes, but now in replay with various other noises. For example, I sampled Huia Echoes joined with the sounds of a New York City cathedral, of the edge of the Arctic Ocean, of Munich, of Germany and of Changsha, China. An unfolding series of 'echoEscapes' — i.e., mixed soundtracks of contemporary spaces fused with the historic Huia Echoes — are aural skeins of intergenerational time as well as of Indigenous and transported, non-Native human and other-than-human languages and voices. These, in turn, can be listened to, repeatedly, in yet other dimensions. It is one of these echoEscapes, composed in Fairbanks, Alaska, that I will be inviting you to share in momentarily.

But, before I do, I want to highlight another aspect vital to this storytelling. Helped along by sources including Linda Tuhiwai Smith's book *Decolonizing Methodologies*, learning a dead birdsong urges learners away from a culture of self-taking credit into a mindset of acknowledgement.[17] I understand this as an intention not only to remember *what* is imparted, but, when possible, to answer *through whom*, with particular attentiveness to holding space for and amplifying those voices who have historically been stifled. Cultivating this consciousness can help in forming reciprocal alliances — new relational networks — characterizing co-creative, re/generative communities, as

---

16 R. Murray Schafer, *The Soundscape: Our Sonic Environment and the Tuning of the World* (Destiny Books, 1993). See also *Voice and Void*, ed. by Thomas Trummer (Ridgefield, CT: The Aldrich Contemporary Art Museum, 2007); Jacob Kirkegaard, 'Aion', *MOMA*, 2013, https://www.moma.org/interactives/exhibitions/2013/soundings/ artists/6/works; and Janet Cardiff and George Bures Miller, 'The Murder of Crows', *Armory*, 2012, http://www.armoryonpark.org/programs_events/detail/ janet_cardiff_george_miller_murder_of_crows

17 Linda Tuhiwai Smith, *Decolonizing Methodologies* (London: Zed Books, 2012). All remaining mistakes and lacks in my own understanding remain my own responsibility.

I continue my decolonization, anti-racist, and other 'learning to listen' homeworks.[18]

In other words, learning to listen also seems to be learning how to participate in healthy community-making, as this continuous project has been from the start. 'Officially' there were no restrictions on use of the Huia imitation recording. To be sure, I checked in with and obtained permission from Cornell Lab of Ornithology's Macaulay Library for permission to work with the soundtrack in above ways. Their staff, particularly Matt Young, have been generous in many ways. In 2016, I took the Lab's field recording course, another help in ear-tuning as well as practice with technologies.

As I continued listening and responsively composing new 'echoEscapes', I became uncomfortable about backgrounding the men on the recording. The fact is, I was far distant from their experiences and relationships and the deep legacies contained in this relict. None of it belonged to me. Moreover, the sacredness of this speaking gift demanded honor. What did that mean? I shared these disquiets with another respected colleague, Princess Daazhraii Johnson, who is Neets'aii Gwich'in, where I lived in interior Alaska.[19] She asked me questions I need to keep asking: how will I keep my engagement with the recording from becoming another act of colonial appropriation? What of consent? What about not only the birds, but the people involved? Reflecting on these led to other callings. While my interest in extinct birds had led me to the soundtrack, listening now opened my ears to human beings I had never heard. Though both men who took part were also deceased,

---

18 For example, see Marilyn Strathern, Jade S. Sasser, Adele Clarke, Ruha Benjamin, Kim Tallbear, Michelle Murphy, Donna Haraway, Yu-Ling Huang, and Chia-Ling Wu, 'Forum on Making Kin Not Population: Reconceiving Generations,' *Feminist Studies*, 45.1 (2019), 159–72. I deeply thank Native Movement, the Gwich'in Steering Committee, and Fairbanks Climate Action Coalition for arrays of workshops, camps, and trainings. Throughout, I am thinking of this definition of 'decolonization': 'The conscious — intelligent, calculated, and active — unlearning and resistance to the forces of colonization that perpetuate the subjugation and exploitation of our minds, bodies, and lands. And it is engaged for the ultimate purpose of overturning the colonial structure and realizing Indigenous liberation.' — Native Movement Training Slide, Summer 2018, https://fairbanksclimateaction.org/events/2018/3/24/decolonize-your-mind-untangle-history-building-liberation

19 Together, we have a co-authored a forthcoming chapter entitled 'To Hope to Become Ancestors', in *What Kind of Ancestor Do You Want to Be?*, ed. by John Hausdoerffer, Brooke Hecht, Kate Cummings, and Melissa Nelson (Chicago, IL: University of Chicago Press, in press).

what responsibilities might my encounters with this historic recording involve to others' ancestors and to their living kin? What might I learn of and from them, were we to find each other, and, if they were willing to share anything, and how? This oriented me within still-multiplying (un)learning paths.

Thanks to initiatives of my colleagues — Drs. Mike Roche, Susan Abasa, and Margaret Forster, experts in historical geography, museum studies, and Māori knowledge and development, respectively — I received an International Visitor Research funding from Massey University. I was granted a month's stay in the former Huia range of the North Island. It was straightforward to find Batley's archives in museums. However, it was difficult to find information concerning Hamana, this itself an expression of settler-colonial legacy. Journalists Kate Evans and Sarah Johnston of Ngā Taonga Sound and Vision kindly introduced me to members of both Batley and Hamana (Salmon) families. I am grateful, sometimes overwhelmingly so, to these families for their generosity in making relations with me in and beyond mutual engagements with the Haman-Huia recording, Huia Echoes.[20]

Before ever meeting Toby Salmon, a great nephew of Henare Hamana's, he had understood that my quest was more than academic. It was also deeply personal, in words he offered — 'a spiritual journey.' During my 2017 visit, Toby brought me and other whānau (extended family) to the gravesite of the late Huia whistler, where he lay next to his wife, Hari. From the small cemetery on a hill, through a grassy-green fog, we overlooked this pair's former home, which had recently collapsed. After ceremonial prayers, we went on a tramp into the same bush into which Toby's uncle had guided in that 1909 search expedition. Under arching canopies of trees and ferns, we listened together.

Finally, before opening into echoEscape.1: 'The Place Where You Go To Listen,' I want to share another integral poem, this one brought to my attention by Dr. Lesley Wheeler, a literary scholar in the United States.

---

20   The Salmon family invited me to return for another visit in February 2020, after this piece went to press. Ngā mihi to Jane and Toby Salmon and Karen and Dave Salmon, on repeat!, for their incredibly generous hosting as we share in such an amazing journey together. It was an all-too brief stay on my part. And, so much remains unsaid here or now that could be said better.

Fig. 3.3 'The Place', Museum of the North, University of Alaska, Fairbanks, Alaska. Photograph by Julianne Lutz Warren (2015).

The words of 'Huia, 1950s' by Hinemoana Baker written in 2004, speak to the experience of a dead language learner:[21]

the huia-trapper

whistles the song
I try to resist

I want to tug
something out of him

the radio voice says
believed to be extinct

Yes, I also feel this need to 'tug', a longing, as I listen to these aural traces. Maybe you will, too.

More questions combine with the others I've mentioned. What is it *I* want to tug out and why try to resist? Might this impulse have

---

21 Hinemoana Baker, 'Huia, 1950s', in *Matuhi/Needle* (Wellington: Perceval Press/ Victoria University Press, 2004), p. 38. I especially thank John Luther Adams, not only for his profound work, but also, as partners — him and Cindy Adams — for good conversations when their lives and those of mine and my partner's have intersected.

something to do with the shapeless loneliness I mentioned above? With anguished hope? With how to love skillfully?

At the same time, what does the complex voice want to tug out of me — that perhaps I should give in to?

Backgrounder for 'echoEscape.1'

Sonic sphere: Naalagiagvik (Inupiaq) 'The Place Where You Go to Listen', a 2006 sound installation by composer John Luther Adams[22]

Location: Unceded Traditional Territories of the lower Tanana Dene Peoples in the Museum of the North, University of Alaska, Fairbanks, Alaska

Geographical Coordinates: Latitude: 64-50'06" N; Longitude: 147-39'11" W

Elevation: 446 feet

Listening date/time: 13 August 2015/9:30am–10:30am

Approximate Nautical Miles from Manawatu Gorge, North Island, Aotearoa, New Zealand, Traditional Region of Rangitāne: 6500

I acknowledge and honor the ancestral and present land stewardship and place-based knowledge of the Indigenous Peoples of Alaska and Aotearoa New Zealand, geographies featured in this echoEscape.

echoEscape.1: 'The Place Where You Go to Listen'[23]
The Approach

I welcome you into echoEscape.1. Please, will you come along into this listening event to practice learning a dead birdsong with me?

---

22   John Luther Adams, *The Place Where You Go to Listen: In Search of an Ecology of Music* (Middletown, CT: Wesleyan University Press, 2009). The title of the work is a literal translation of the Inupiaq place name 'Naalagiagvik'. Listening samples are available here: https://www.uaf.edu/museum/exhibits/galleries/the-place-where-you-go-to/

23   In various interactions, this essay also has taken forms as a sound art piece with great thanks to Mickey Houlihan (curveblue.com) for audio editing and mixing and to Joe Shepard for audio restoration and editing..

This August day, the clouds are heavy and low — swells of grey down.

The sky has just rained, darkening the pavement. Water is dripping from green boughs of spruce and birch branches. An easy wind smells like wet grass.

Downhill, in a nearby grain-field turned gold, Sandhill Cranes are chiming — 'deldal' in Benhti Kanaga, a local Dene language, 'a crane is calling'[24] — in the few days left before migration.

The moon will be new tomorrow. The tiniest sliver of today's waning crescent, though invisible, arcs slowly above the horizon.

The morning sun's light, filtered through the blanket overhead, is muted. A listener might want to wear a sweater.

Once inside the museum, a listener hears kids' laughter ringing off slick floors and glass doors. Heels click past the gift shop, then climb a staircase to the second floor with a bay of windows on the right. To the left, there is a gallery with an alcove.

A sign on a door in the alcove says, 'enter quietly'.

A listener may want to pause before doing so. Turn around to look, again, through the wide span of glass. Generations of Dene have come to Troth Yeddha', or, 'the ridge of the wild potatoes', to harvest these legumes. With such a sweeping vantage, this hill has long been an attractive meeting site. Outside, beyond today's rain veil, are mountains far older even than this Land's First Peoples. The immensity of that rock, though over a hundred-miles distant, feels demanding. On a clear day, the 'Three Sisters' are eye-magnets — their snowy tops gleam gold-pink. And, more to the right, to the southwest, would be Denali — the tallest North American peak.

Unless, perhaps, this summit has fallen during the night. It can't be known for sure, unless the clouds withdraw, whether or not the mountain remains standing at all, can it?

An introductory plaque on the alcove door describes 'The Place Where You Go to Listen'. It tells how an Iñupiaq legend was in the memory of a composer, John Luther Adams, who imagined and, with the help of others, built this sound art installation. A woman, so the legend says, sat quietly in a place called 'Naalagiagvik' on Alaska's

---

24 The final 'l' is a voiceless fricative. *Lower Tanana Athabaskan Dictionary*, compiled and edited by James Kari, Alaska Native Language Center, First Preliminary Draft, 1994, https://uafanlc.alaska.edu/Online/TNMN981K1994b/kari-1994-lower_tanana_dictionaryf.2.pdf

Arctic Coast. In that place, she — her name has not been given — heard things.

In contrast to the open Arctic plain spreading into a vast, salt-smelling sea, the room a listener is about to enter is a close space — about ten by twenty feet, with no windows. There is one long wooden bench in the center. Floors, ceiling and walls are white, except for one wall that glows with the only source of light. This light slowly, barely perceptibly, changes color — in summer, of yellows and green-blues — seasonal, circadian hues tune with the unceasing flow of noise vibrating from a surround of speakers. The noise in the speakers emanates from machines that translate sources of real time physical conditions outside into sounds — filtered, tuned and tempered — merged into a continuous electronic stream. This resonating stream, fluid with emergent tones and rhythms, is a polyphony of irregular seismic groans at foot-shuddering level, at ear-height, the reliable voice of moon — from the perspective of Earth's horizon — waxing and waning, rising and setting with the chorusing sun sung through sound-damping prisms of palls and mists of air. From ceiling speakers, when aurora are active, fluxing bells tinkle down voicing them.

Before Entering

Be warned: Upon first opening the door, a listener sometimes feels repelled by the room's chaotic acoustic atmosphere, even afraid. In past visits, I have hesitated on this threshold, my heart pounding. From inside I have watched many would-be listeners crack open the door, then, retreat. But, please, won't you enter, with me, together?

One of the most important things about this place is that a listener *may* go. Anyone who enters this room can exit the same way.

But, why not stay for awhile, perhaps, learning 'to be still?' Although neither you nor I may yet understand what we carry inside — Huia Echoes — is a performance of remembering.

Crossing the Threshold, Entering

This place, says Adams, 'is not complete until you are present and listening'.

First, a listener must resist leaving. Second, a listener will do well *not* to resist giving in before, at some point, departing.

A listener clicks the open the door. On one side of the threshold, reverberating footsteps, adult talk and pealing laughter. On the other side, the continually streaming noise of The Place.

The door clicks shut behind us. Now, hear only the noise, streaming into a listener's gut, the rocking Earth rumbles. Shhhhh — recall, the new moon, present, yet silent. Tone-rays of sun lap a listener's cheeks. When clouds thin outside, an aural prism releases daylight's chorus. Billows thickening sounds a shaded indoor quiet. Tides of tinkling auroral bells wash in and out over a listener's head.

Ears, set loose, float away, drifting birds in unceasing, ever-changing currents of sound.

If this is a dream of voices dreaming. Who is the dreamer? The whole Earth? The human composer? A present listener? Past and future ears? Some other source? Is it possible for all to dream the same dream, and differences?

A listener doesn't know.

A Māori chant shared in the nineteenth century by Te Kohuora of Rongoroa, speaks, Na te — that is, from the — primary source, rising-thought-memory-mind/heart-desire...[25]

Why not wonder...If

when the mind/heart remembers the source and desires what emerges, isn't that supreme hope? Who can solve hope?

Te Kohuora's chant continues, *Te kore te rawea* — or, say, unbound nothingness — stirring hau — breath of life and growth — moving through darkness, the world, the sky, moon, sun, light — day! — earth (female) — and — sky (male) — and ocean, the children of earth and sky, food plants, forests, lakes and rivers, ancestors of fish, lizards, birds...life and death and life...

In Te Reo Māori, change happens across the pae — liminal spaces of potential emergence — between life and death, light and dark, silence

---

25 From 'creation chant' quoted in Anne Salmond, *Tears of Rangi: Experiments across Worlds* (Auckland: Auckland University Press, 2017), pp. 11, 14.

and noise, absence and presence, inhabitants and visitors, one kind of being and another.[26]

In The Place, in this room, imagine many pae — the sound is never-ending and never the same, unfolding in time. There are low tones and high, consonances and dissonances begin and end, the inaudible outside is audible inside mingled with dim ambience and subtly changing colored light. The door that opened will close. The listener who entered —   *inhales   exhales*   will eventually leave.

Replaying Huia Echoes[27]

Now a listener presses a button on a play-back machine releasing the legacy of interrelations voiced by Huia Echoes, foreign to this place, into the many other exchanges already happening: Two men — now both dead — the colonized one sharing a taonga — a treasured thing — with the colonizer, a colonizer, working to de-colonize, passing it on, asking — who am I? Who are you? Listening. The taonga is the imitation of two birds reciprocating want, dead voices sound animate. What is breathless, and unchanging, sings, full of breath, if not changing, bringing changes (or, were the multiple voices always ever asleep, now waking?)

The room's noise never stops. It courses in eddies and flushes downstream.

> While, Huia Echoes tracks a round channel.

As the noise of this inside world is ever-changing as is the outside one, Huia Echoes is a recording with a chorus of voices on repeat — beginning-middle-end — the same, over and over again.

> Or   are they?

What happens across the pae between this room of noise and the machine-saved echoes of lost man-and-bird?

---

26   Pae (threshold, middle ground) in Salmond, *Tears of Rangi*, p. 3.
27   The looped recording can be heard here: https://merwinconservancy.org/wp-content/uploads/2015/11/HuiaEchoesLoopCornellTrack_1.wav (part of my earlier essay https://merwinconservancy.org/2015/11/the-poetry-lab-hopes-echo-by-author-julianne-warren-center-for-humans-and-nature/).

Huia Echoes sounds mingled and disruptive, tender and insistent. Echoes echo. Their voices are buoyed by waves   sink, and soar. The bird-man chorus varies with the outside weather heard inside the world of The Place. While, auroral bells curtain the recording's higher pitches, making them hard to distinguish. When the clouds disperse, beyond the walls, the octaves of the sun's choir widen and brighten. The sun brightening, paradoxically, shades Huia Echoes' tones. The clouds, darkening, unexpectedly, expose the re-playing voices. Aurora, quieted, release the higher notes of extinct bird-man songs. Catching a sonic wave — and, sometimes don't you hear your own speech — voices, caught, sway up and down — male and female oscillate:

> *Here I am.*
> > *Here I am.*
> *Over here.*
> > *Here am I.*
> Who are you?   Who I am?   Am *I* you?   You, me?
> Are we 'we'?

In the stream of sounds, the calls of Huia Echoes alter. Or, is it Huia Echoes — as leaves do the wind — re-directing the room's flow of sounds? Are the ears of a listener transformed? Who is the changer, and, who being changed, in this dream of dreamers dreaming? Who or what has will? What is trust? What desire?

Huia Echoes Stopped

After listening awhile, a listener turns off the machine replaying the bird-man recording. Feel the relief in the silence of their cluttering tones within this room full already overflowing with sounds. Is that relief a germ of forgetting?

Yet, having grown familiar with the added presence of Huia Echoes, the room now also sounds wrong. Feel the absence of the composite voice — with a pang — as a plangent void.   Is that pang a seed of remembering?

But,   what if   Huia Echoes, rather than remain a captive of the machine, found a sonic   wormhole   and escaped?

Though the recording is off, catch those phrases of the bird-man voice repeating

— a ghost of tones —

Is that audible phantom, my desire or Echoes'?

A Listener Plays

Still captivated by the ever-rolling sounds of the room, a listener stays on, hearing, but not still, puzzling, a contribution.

A listener's voice turns on. Wells up, though it does not want to shout nor need to cry out, but, somehow, at the same time to dissolve yet   stand out.

A childhood's nursery rhymes labor as breath is tugged out of a listener's mouth while another listener can bear other words by their tongues who play wordplay words *Three blind mice* repeats, mantra-like   syllables   hamper   and scamper,   unpredictably *see how*

they run?

How much wood can a woodchuck chuck? ᵐᵘᶜʰ *wood could the woodchuck chuck* an adverb ˡᵉᵃᵖˢ into the drink.

PeterPiperpickedapeckofpickledpeppers tumbles out next   and   cannot outrun time not even backward *pepperspickledofpeckpickedPeterPiper* as voiceless plosives resist   flowing

Humpty Dumpty's manifest consonants spill out — *and all the king's men couldn't put Humpty Dumpty together again* as miscarried   skipping stones,

plunking into this ocean of sound   Whereas,

Mary   had   a   little   lamb catches   a   current   resonates and   clashes   randomly in the swirls   until *the lamb was sure to* goOOOOOOOoooooooo blends with sun's hum tuned in a human listener's capacity to a fundamental frequency of the daily spin of Earth — twenty-four point two seven hertz is a tempered G – eooooooooo the tiny vowel – ooooooooo pools into a quiet layer of sound an octave between the organ of solar light and shimmering auroral bells harmonizing with the whole planet

A listener becomes a self-conscious composer

A coda happens

three blind mice   three blind mice   three blind mice   see how they run? bends the old rhyme into a question

hum goOOOOOOooooooo......   Wait...
   Run, where?   Blinded, how?   Why see?   Why listen?
Why sing?

Crossing the Threshold, Exiting

A listener gets ready to leave 'The Place', where it has been safe both to give in and to resist, be distinct. One of the most important things about this room is that the listener may find a way out.

A listener clicks open the door. On one side, the continually streaming noise of The Place. On the other side, reverberating footsteps, reverberating talk and peals of laughter.

The door clicks shut behind a listener. Heels click along the shiny tile floor, past the bay of windows, to the left. A listener pauses, again, to look out.

The moon is still new, invisible as the sun — audibly chorusing back inside — still hidden behind silver billows, as are the distant mountains.   A listener has kept faith in them.

A listener's footsteps now ring down the stairs, past the gift shop, through the glass doors.

Outside

A listener does not hear deldal once back outside — that is, the cranes are no longer chiming. Beyond the still-billowing clouds, a fingering menace is closing around Earth. The menace twists Earthlings into an undying sameness, a dry, pae-less monotone that repeats, *'believed to be extinct.'*

But, still, listen — there is a courage of breath that can say *yes!* and also *no!* And, I don't know! And, in between, spaces of potential emergence from which many voices might echo.   A dead birdsong learner, too, may join in, also unlearning what has never been good.

And, what if Huia Echoes did pass through a wormhole back there in the chaos of sounds? What if they did   refuse the machine, and anyone's will to turn them off or on?

In the lingering ear-tuning of The Place, it's not only that a listener hears music in the whir of a fan, or mistakes a distant chainsaw cutting firewood as Dene languages more-than-surviving, there also is something going on with the birds.

The Birds

It is early spring, Huia Echoes wakes me from sleep. They are singing in the forest beyond an open window.   I leap out of bed   fumble, bleary-eyed, lean out, ears tune in and hear

 those Swainson's thrushes

  as I suddenly recognize them.[28]

They had picked up the lost bird-man voice.

 What if these boreal flesh-and-feather throats echoing the echoes of extinct birds from a world away?

A listener cherishes that dream all the more as a/wake   a persistent, ringing void   keeping   shalak naii (all our relations)[29]

 in the world-alive

  in play.[30]

Bibliography

Abram, David, *Becoming Animal: An Earthly Cosmology* (New York, NY: Vintage, 2011).

Adams, John Luther, *The Place Where You Go to Listen: In Search of an Ecology of Music* (Middletown, CT: Wesleyan University Press, 2009).

Anderson, Atholl, 'A Fragile Plenty: Pre-European Māori and the New Zealand Environment', in *Environmental Histories of New Zealand*, ed. by Eric Pawson and Tom Brooking (Oxford: Oxford University Press, 2002), pp. 35–51.

---

28 A recording of Swainson's Thurshes can be heard here: https://search.macaulaylibrary.org/catalog?taxonCode=swathr&mediaType=a&q=Swainson%27s%20Thrush%20-%20Catharus%20ustulatus

29 In Gwich'in, another Arctic Athabascan language.

30 This song offers a beautiful tribute to Huia and relations: Maisey Rika, 'Reconnect,' on *Maisey Rika* (Moonlight Sounds, 2009). Kia ora (Special thanks), Toby and Jane Salmon — 'Ko matou nga Kaitiaki mo apopo.'

Baker, Hinemoana, 'Huia, 1950s', in *Matuhi/Needle*, Hinemoana Baker (Wellington: Perceval Press/Victoria University Press, 2004), p. 38.

Batley, R. A. L. Archives, Box 2011.117.1 MS 177, Whanganui Regional Museum, https://macaulaylibrary.org/asset/16209

Buller, Walter L., *A History of the Birds of New Zealand, Vol 1* (London: self-published, 1888).

Cain, Trudie, Ella Kahu, and Richard Shaw, eds, *Turangawaewae: Identity and Belonging in Aotearoa New Zealand* (Auckland: Massey University Press, 2017).

Campbell, E. E., *Central District Times*, Letter to the Editor, 3 December 1974 in RAL Batley, Archives Box 16, 2011.117.1, Whanganui Museum, NZ.

Cardiff, Janet, and George Bures Miller, 'The Murder of Crows', Armory, 2012, http://www.armoryonpark.org/programs_events/detail/janet_cardiff_george_miller_murder_of_crows

Forster, Margaret E., 'Recognising Indigenous Environmental Interests: Lessons form Aotearoa New Zealand,' Presented at *International Political Science Association, Research Committee 14: Politics and Ethnicity. The Politics of Indigenous Identity: National and International Perspectives*. Dunmore Lang College, Macquarie University, Sydney (July, 2013).

Jensen, Derrick, 'Beyond Hope', *Orion Magazine*, 2 May 2006, https://orionmagazine.org/article/beyond-hope

Kirkegaard, Jacob, 'Aion', *MOMA*, 2013, https://www.moma.org/interactives/exhibitions/2013/soundings/artists/6/works

Knight, Catherine, *Ravaged Beauty: An Environmental History of the Manawatu* (Ashhhurst: Totara Press, 2014).

Licht, Alan, 'Sound Art: Origins, Development and Ambiguities', *Organised Sound*, 14.1 (2009), 3–10, https://doi.org/10.1017/s1355771809000028

Merwin, W. S., 'Learning a Dead Language', in *Migration: New and Selected Poems*, W. S. Merwin (Port Townsend, WA: Copper Canyon Press, 2005), p. 41.

Norman, Geoff, *Bird Stories: A History of the Birds of New Zealand* (Nelson: Potton and Burton, 2018).

Orbell, Margaret, *The Natural World of the Māori* (Auckland: Collins, 1985).

Phillips, W. J., *The Book of the Huia* (Christchurch: Whitcombe and Tombs Ltd, 1963).

— 'Huia Research: 1927–1954, (MU000235)', Dominion Museum (creating agency) (Wellington: Museum of New Zealand/Te Papa Tongarewa, [n.d.]).

Rika, Maisey, 'Reconnect', in *Maisey Rika* (Auckland: Moonlight Sounds, 2009).

Riley, Murdoch, *Māori Bird Lore* (Paraparaumu, NZ: Viking Sevenseas, NZ Ltd, 2001).

Salmond, Anne, *Tears of Rangi: Experiments across Worlds* (Auckland: University of Auckland Press, 2017).

Schafer, R. Murray, *The Soundscape: Our Sonic Environment and the Tuning of the World* (Destiny Books, 1993).

Selby, Rachel, Malcolm Mulholland, and Pataka Moore, eds, *Maori and the Environment* (Wellington: Huia Publishers, 2010);

Smith, Linda Tuhiwai, *Decolonizing Methodologies* (London: Zed Books, 2012).

Szabo, Michael, 'Huia, The Sacred Bird', *New Zealand Geographic*, 20, October-December (1993), https://www.nzgeo.com/stories/huia-the-sacred-bird

Lower Tanana Athabaskan Dictionary, compiled and edited by James Kari, Alaska Native Language Center, First Preliminary Draft, 1994, https://uafanlc.alaska.edu/Online/TNMN981K1994b/kari-1994-lower_tanana_dictionaryf.2.pdf

Tennyson, Alan, and Paul Martinson, *Extinct Birds of New Zealand* (Wellington: Te Papa Press, 2007).

Trummer, Thomas, ed., *Voice and Void* (Ridgefield, CT: The Aldrich Contemporary Art Museum, 2007).

Warren, Julianne, 'Hopes Echo', The Poetry Lab, *The Merwin Conservancy*, 2 November 2015, https://merwinconservancy.org/2015/11/the-poetry-lab-hopes-echo-by-author-julianne-warren-center-for-humans-and-nature

Warren, Julianne, and Daazhraii Johnson, 'To Hope to Become Ancestors', in *What Kind of Ancestor Do You Want to Be?*, ed. by John Hausdoerffer, Brooke Hecht, Kate Cummings, and Melissa Nelson (Chicago, IL: University of Chicago Press, in press).

Whatahoro, H. T., *The Lore of the Whare-Wānaga: Or Teachings of the Māori College on Religion, Cosmogony, and History Vol. 1: Te Kauwae-Runga, or 'Things Celestial'*, trans. by S. Percy Smith (Cambridge, UK: Cambridge University Press, 2011), https://doi.org/10.1017/CBO9781139109277

3. Humilities, Animalities, and Self-Actualizations in a Living Earth Community

Paul Waldau

In preparation for the workshop's dialogue on 'multiple ways of being and knowing' in our shared 'living Earth community', I attempted to ascertain themes relating to the following question: *how might an individual today choose actions that celebrate the plain fact that each of us is a member of a species that has only sometimes, and not often lately, been a responsible member of the Earth community?* I present my findings in this exploratory piece.

My framing of these issues focuses particularly on the importance of different forms of humility. I suggest that different forms of humility are needed because each of us is a member of human-centered communities that have, whether intentionally or not, produced diverse harms beyond the species line that many individuals within our own species and, in particular, the major institutions of modern industrialized societies have long *celebrated* rather than condemned.

My framing also foregrounds our obvious animality, although again I want to spur my own thinking by using the plural 'animalities', since lives on this planet are unbelievably diverse and always embedded in a more-than-animal context. I refer here both to those nonhuman lives we name with words like 'plant' and phrases such as 'the material world' to denote those parts of the universe that our host culture overwhelmingly treats as non-living, and thus merely a resource for our use and benefit.

My experience over the last half-century has suggested to me that no rich form of 'self-actualization' is possible for us when humans

    https://doi.org/10.11647/OBP.0186.03

claim to be separate and superior, as occurs habitually through the demarcating property of language that produces categories such as 'humans and animals'. I take human exceptionalism to be the dominant narrative of our time, even though in our received wisdom traditions there are many profound formulations about recognizing the importance of both human and nonhuman 'others' whenever any human individual or group seeks full self-actualization.[1] I offer here a few forthright statements that make plain the importance of such wisdom. The first is from Viktor Frankl.

> [S]elf-actualization is possible only as a side-effect of self-transcendence.[2]

A pair of comments from Thomas Berry takes the issue well beyond the species line:

> [W]e must say that the universe is a communion of subjects rather than a collection of objects.
>
> Indeed we cannot be truly ourselves in any adequate manner without all our companion beings throughout the earth. The larger community constitutes our greater self.[3]

Beware Bootlegging. I also use the plural 'self-actualizations' in this chapter because I intentionally want to call out another issue — it does not follow that one's own notions and/or attempts at self-actualization provide any sort of paradigm by which the self-actualization of other animals, whether human or not, can be measured. Instead, I go forward on the assumption that, in any group (and this gathering of chapters

---

1   I have previously defined 'human exceptionalism' in my book *Animal Studies — An Introduction* (Oxford: Oxford University Press, 2013), as follows (p. 8): 'Human exceptionalism is the claim that humans are, merely by virtue of their species membership, so qualitatively different from any and all other forms of life that humans rightfully enjoy privileges over all of the earth's other life forms. Such exceptionalist claims are well described by [James] Rachels as "the basic idea" that "human life is regarded as sacred, or at least as having a special importance" such that "non-human life" not only does not deserve "the same degree of moral protection" as humans, but has "no moral standing at all" whenever human privilege is at stake'.

2   Viktor E. Frankl, *Man's Search for Meaning: An Introduction to Logotherapy*, 4th edn (Boston, MA: Beacon Press, 1992), p. 115.

3   The second quote is from Thomas Berry, 'Loneliness and Presence', in *A Communion of Subjects: Animals in Religion, Science, and Ethics*, ed. by Paul Waldau and Kimberly Patton (New York, NY: Columbia University Press, 2006), pp. 5–10 (p. 5). The first quote was said by Berry on multiple occasions, and it appears at p. 7 of the same collection.

would provide a paradigmatic example of the following), there will be different forms of self-actualization. One widely successful form appears in service traditions, and yet other forms appear in meditation traditions. Many other forms appear in instances where individual humans have found a way to stand outside the penchant for self-preoccupation that individuals in our own species so often exhibit. In such instances, these individuals have *thereby approached particularly fulsome forms of self-actualization.*

Based on the personal and communal experiences that have led me to describe issues as I do above, and based on the challenges I tried to meet in my previous book-length projects (both single-author publications and the two edited collections *A Communion of Subjects* and *An Elephant in the Room*),[4] I am currently finishing a book that will carry the title *The Animal Invitation: Science, Ethics, Religion and Law in a More-Than-Human World.* This book is an attempt to say what five different human domains — science, ethics, religion, law, and education — might look like if we took our animality seriously.

To introduce the issue further, I include next the opening two paragraphs of the book, after which I will try to sketch out ways in which I think each of the four eminently human efforts described in the subtitle — science, ethics, religion, and law — must always be living efforts (this claim, which is by no means novel in regard to any of these four domains, is related to how I discuss our own animality throughout the book). In my closing comments below, I will address both formal and informal education, since this theme is a meta-topic, as it were, of the chapters addressing science, ethics, religion and law.

> Animals invite us. This world-constituting fact is true whether we are talking about humans inviting humans, or, the focus of this book, nonhumans inviting human awareness, co-existence, appreciation, and even awe. One domain after another of our human existence, including often our daily lives, reveals the astonishing variety and depth of these invitations.
>
> It is both of these features — variety *and* depth — that are, tellingly, reflected in the human domains we know as 'science,' 'ethics,' 'religion' and 'law.' Admittedly, the great variety of approaches, which has

---

4   *An Elephant in the Room: The Science and Well-Being of Elephants in Captivity*, ed. by Debra L. Forthman, Lisa F. Kane, David Hancocks, and Paul Waldau (North Grafton, MA: Center for Animals and Public Policy, 2008).

spawned many different ways of talking and thinking about the animal invitation, reflects both deep acknowledgements and facile dismissals. Considered alone, the variety is revealing, for it reflects basic features, especially the finitudes, of our human capacities. But it is the depth evident in many humans' recognition of the animal invitation which, though less commonly encountered than diversity, reflects best the fecundity and vivifying power of human thinking and action. As this book will show, human possibilities, narrow and broad, play out in the depth and variety of responses to the animal invitation that are evident in different human groups' claims of identity, community, compassion, awareness, self-delusion, self-inflicted ignorance, and so much more.

In the following four sections, I raise the issue of whether our astonishing achievements in science, ethics, religion, and law are (i) helpfully seen as *eminently animal achievements,* and (ii) better understood when each of these four domains is discussed primarily as an ongoing commitment of our kind of animal that *must* be understood and experienced as 'living now', rather than 'eternally fixed' or as an 'absolute truth'. Correspondingly, treatment of any of these domains as irrevocably fixed defeats what can be thought of as the vivifying and enabling genius of each of these living domains as a human achievement. I suggest in the book, then, that it takes truly living, responsive forms of each of these human achievements to move individual humans in the direction of full actualization of our human animality.

Human Science in a More-Than-Human World

That our sciences have organic features is strongly hinted in the long history of shifts in ideas and changes in governing paradigms across the centuries. Organic features of many sciences are also seen in the unbelievable rate of new discoveries in recent decades, for these discoveries have produced shifts in particular scientific communities' dominant ways of thinking. I want to add, though, that it remains my impression (perhaps a result of my 'education') that the western science tradition in some ways still does not feature 'living aspects' quite as fully as do ethics, religion, and law.

Human Ethics in a More-Than-Human World

I'm only too aware that ethics has long been taught in the western intellectual tradition as a set of answers to questions such as 'what is the right thing to do?' and 'what does it mean to be a moral and/or good person?' Having taught ethics now for over twenty years, I do not think such formulations are helpful, nor do I think these formulations reflect that ethics is, and always needs to be, *very alive indeed*. For this reason, I have come to see such views of ethics as a failure to detect the true heartbeat that takes place as we embrace, develop, and seek full actualization of our human ethical abilities. A question that does prompt us to hear more clearly the heartbeat of our ethical abilities is what I have come to call 'the root question' of ethics, namely, 'who are the others?' This is an abbreviated version of what is, in our daily lives, a far more complicated rendition of this root question, which can be stated in a variety of ways — here's one version that I think captures some of the animal and human genius of the abilities we call 'ethics': 'Who are the others about whom I should care *given that I have finite abilities and there are, as a practical matter, many other limits on my ability to care*?'

The principal point in the book's chapter on ethics is that such root questions, and of course the abilities that we use in pursuing our own answer, reflect what can only be described as *eminently animal abilities*. I do not mean to suggest with 'eminently' in the prior sentence that each and every kind of animal features the high-level abilities we call ethics — my guess is that only some animals do (caring about 'others' is more common, I suspect, in mammals, but there is much to suggest that some birds and a variety of non-mammals also have some feature in their life that, in effect, can be described as a version of the 'who are the others?' question).[5]

---

5   Note, for example, the great range and diversity of life explored in the following titles: Frans de Waal, *Are We Smart Enough to Know How Smart Animals Are?* (New York, NY: W. W. Norton, 2016); Robin Wall Kimmerer, *Braiding Sweetgrass: Indigenous Wisdom, Scientific Knowledge, and the Teaching of Plants* (Minneapolis, MN: Milkweed, 2013); David Haskell, *The Songs of Trees: Stories from Nature's Great Connectors* (New York, NY: Penguin, 2017); David Abram, *Becoming Animal: An Essay on Wonder* (New York, NY: Pantheon, 2010); M. D. Olmert, *Made for Each Other: The Biology of the Human-Animal Bond* (Cambridge, MA: Da Capo Press, 2009); Jonathan Balcombe, *What a Fish Knows: The Inner Lives of Our Underwater Cousins* (New York, NY: Scientific American/Farrar, Straus, and Giroux, 2016); Jennifer Ackerman, *The*

In this chapter, I suggest that one *cannot* understand an ability of the ethical sort without affirming that ability's animal origin and nature. As a segue into the following comments about humans' spiritual/religious awarenesses, let me add that I have come, after a half-century of studying religious traditions, to think that much, if not *all*, of great value in our religious traditions follows from the eminently animal nature of ethics. Religious traditions are of particular interest to me on account of their following aspects: the role of narrative; the pervasive degree to which our worlds feature sacredness and gift in connection with real places and other living beings; and the insightful observation that relational epistemologies are crucial to each of us recognizing much of who we are.

Human Religion in a More-Than-Human World

Here I tread on sensitive ground — I do this intentionally *and reverentially*, recognizing that there is no single definition of religion that I might employ to argue that 'religion must be alive in order help humans self-actualize.'

As I get older, I'm less inclined to preface the following claim with *mea culpa*, but perhaps I should as a way to underscore my theme of 'humilities' — much that is called 'religion' fails to help 'adherents' or 'believers' self-actualize (in the sense I use this term in these short comments). Yet our spiritual/religious domains seem to me, after a half-century of immersion in studying religion, to include a great variety of options, some of which embrace responsibly rather than repudiate what it means for religious awareness to respect and nurture our many animal-based abilities, finitudes, fragilities, organic births, decline, and eventual death. Religious awareness, when it acknowledges our animality in responsible, foundational ways, will itself be *truly alive and living in every sense* that I am an animal now alive and living.

Genius of Birds (New York, NY: Penguin, 2016); Marc Bekoff, *The Emotional Lives of Animals: A Leading Scientist Explores Animal Joy, Sorrow, and Empathy, and Why They Matter* (Novato, CA: New World Library, 2007); Neil Shubin, *Your Inner Fish: A Journey into the 3.5-Billion-Year History of the Human Body* (New York, NY: Pantheon Books, 2008).

Human Law in a More-Than-Human World

Law (by which I mean 'legal systems', of which there are at least seven distinct major traditions and obviously many different minor variations) may also seem, like science, somewhat of a challenge to fit into the 'living' paradigm. Yet any study of comparative law makes it obvious how fully *constructed* each individual legal system is, and how such 'construction' has features that are easily discerned to be 'living', in the sense that I'm using that broad term in this short paper. This can be observed in these two comments by Robert Cover:

> To live in a legal world requires that one know not only the precepts, but also their connections to possible and plausible states of affairs.[6]
> Law is the projection of an imagined future upon reality.[7]

The need for stability in legal systems, especially as they are part of complex societies, creates features and pressures that tend to make legal systems 'conservative', 'predictable', and subject to forces that easily and often have made enactment and enforcement of 'law' the prerogative of reactionary forces.[8] Consider the exclusion implied in Cicero's seemingly inclusive comment that 'we are all servants of the laws, for the very purpose of being able to be freemen.'[9] The 'we' today might seem to refer to the human group alone, but Cicero through this claim in actuality hides two recurring facts. Human groups now use, and seemingly forever have used, 'law' (developed legal systems) to subordinate not only nonhuman animals and the more-than-human Earth, but also marginalized, politically powerless human groups.

Although the contemporary movement widely known as 'animal law' has for the last two decades challenged such a narrow construction of law, public policy circles today nonetheless remain ignorant of and

---

6   Robert M. Cover, 'The Supreme Court, 1982 Term – Foreword: Nomos and Narrative', *Harvard Law Review*, 97.4 (1983), 4–68, at 10.

7   Robert M. Cover, 'Violence and the Word', *Yale Law Journal*, 95.8 (1986), 1601–29, at 1604, https://doi.org/10.2307/796468

8   It should be noted that my generalizations here do not apply to Indigenous legal systems.

9   Cicero makes this comment in 'The Speech of M. T. Cicero in Defence of Aulus Cluentius Habitus' (M. T. Cicero, *The Orations of Marcus Tullius Cicero*, trans. by C. D. Yonge (London: Henry G. Bohn, 1856), Perseus Digital Library, http://www.perseus.tufts.edu/hopper/text?doc=Perseus%3Atext%3A1999.02.0019%3Atext%3DClu). See paragraph LIII, paragraph 146.

unconcerned about the 'animal question'. There are changes afoot today by which the living features of law can be seen, but since, characteristically, 'the political trumps the legal', the full potential of public policy for the more-than-human world remains, as of yet, unrealized.

Some Final Comments on Human Education

The education theme is, as noted above, a meta-theme in the forthcoming book. In my *Animal Studies — An Introduction*, I worked with both formal and informal education, both of which are encapsulated by an observation made by the English philosopher Stephen Clark: 'one's ethical, as well as one's ontological framework is determined by what entities one is prepared to notice or take seriously'.[10] I entered the academic world because, for me, it *is* a place a daring, and so much so that, at its best, the academic world fosters critical thinking that allows for self-criticism along the lines of Theodore Roszak's 'But then let us admit that the academy has very rarely been a place of daring'.[11] David Orr adds a further dimension to this discussion, extending the issue across the species line — 'The truth is that without significant precautions, education can equip people merely to be more effective vandals of the earth'.[12]

One way in which our society has been equipping 'educated humans' to be 'effective vandals' (or, in Aldo Leopold's phrasing, 'conqueror of the land-community' rather than 'plain member and citizen of it') is categorical division of humans from other animals.[13] This framing defeats us even as it prompts ignorance that leads to great harms to other animals and their local communities. Teachers and students who insist on language that foregrounds a 'human/animal dualism' seem to me to have less chance, often none at all, of accurately assessing themselves or counseling other humans in ways that lead to greater prospects of self-actualization. Why do I suggest this? Because our evident mammality,

---

10   Stephen Clark, *The Moral Status of Animals* (Oxford: Clarendon, 1977), p. 7.

11   Theodore Roszak, 'On Academic Delinquency', in *The Dissenting Academy*, ed. by Theodore Roszak (New York, NY: Vintage, 1968), pp. 3–42, at 4.

12   David Orr, *Earth in Mind: On Education, Environment, and the Human Prospect* (Washington, DC: Island Press, 1994), p. 5.

13   Aldo Leopold, *A Sand County Almanac, with Essays on Conservation from Round River* (New York, NY: Ballantine, 1991), p. 240.

primatehood, and ape-ness are radically (that is, at the root) denied by the dualism.

A key feature of our local formal education — the two-part division of 'higher education' into the sciences, on the one hand, and the 'arts and humanities', on the other — continues to foster the notion that human possibilities are the paradigm of achievement for any living being. In effect, the two-part university has features that legitimize human exceptionalism in a more-than-human world — this is one way that it equips us to be effective vandals of our shared world. Moreover, education further vandalizes in those areas of formal education *where the ideology 'all humans matter' inadvertently masks harms done to many human animals as well*. Thus, in the book, I suggest that teaching about science, ethics, religion, and law in virtually all mainline institutions today presents a face of human exceptionalism that goes beyond harms to nonhumans and their communities because, ironically, formal education in practice continues to hold in place the privilege of only some humans.

A Near-Term Task

I have come to think of our personal and social tasks as finding ways to re- assert our animality, even though these fundamental features of our lives are hidden in plain sight, as it were. These animal abilities are, I suggest, the very condition of our (i) doing science thoroughly and effectively, (ii) pursuing 'living' forms of ethics, (iii) fostering diverse opportunities for spiritual and religious awareness that are truly alive and free, and (iv) creating legal systems that create and project for ourselves a future of responsible membership in the larger community.

A Longer-Term Task

My sense that we can do such work by returning to a full, gracious acknowledgment of our own animality needs, I think, to be supplemented by affirmations of the fact that 'our larger community' includes more than animals alone — insights about the plant world are cascading into our awareness again by virtue of creative scientific work, and our connection to the whole earth is, of course, something that many small-scale cultures have long known. The senses of 'gift' and 'community' found in writers such as Robin Wall Kimmerer, Richard Wagamese, and Linda Hogan reveal that our human forebears knew a great deal about setting the stage for the emergence of a larger community and for forms of self-transcendence that such a community offers, and thereby helps make our own self-actualization possible and fuller.

Let me end on notes that are intentionally provocative and personal — I have come to think of denials of humans' evident animality as cowardice in the face of reality. I am an animal, and so are members of my human community. I love them not in spite of their animality, but because of their animality. And I have come to recognize that I cannot 'know myself', nor it seems to me can any human come to know the possibilities of their life well, without coming to terms with the plain fact that *we are now and have always been and will always be animals*. By acknowledging our animality, we stand to open up key possibilities for self-actualization. This is why my forthcoming book, as well as the present volume in which this chapter appears, attempts to explore our scientific, ethical, religious, and social sensibilities that permit forms of life and a rule of law that are fair to all members of our extended, larger community.

Bibliography

Abram, David, *Becoming Animal: An Essay on Wonder* (New York, NY: Pantheon, 2010).

Ackerman, Jennifer, *The Genius of Birds* (New York, NY: Penguin, 2016).

Balcombe, Jonathan, *What a Fish Knows: The Inner Lives of Our Underwater Cousins* (New York, NY: Scientific American/Farrar, Straus and Giroux, 2016).

Bekoff, Marc, *The Emotional Lives of Animals: A Leading Scientist Explores Animal Joy, Sorrow, and Empathy, and Why They Matter* (Novato, CA: New World Library, 2007).

Berry, Thomas, 'Loneliness and Presence', in *A Communion of Subjects: Animals in Religion, Science, and Ethics*, ed. by Paul Waldau and Kimberly Patton (New York, NY: Columbia University Press, 2006), pp. 5–10.

Cicero, M. T., *The Orations of Marcus Tullius Cicero*, trans. by C. D. Yonge (London: Henry G. Bohn, 1856), Perseus Digital Library, http://www.perseus.tufts.edu/hopper/text?doc=Perseus%3Atext%3A1999.02.0019%3Atext%3DClu

Clark, Stephen, *The Moral Status of Animals* (Oxford: Clarendon, 1977).

Cover, Robert M., 'The Supreme Court, 1982 Term – Forward: Nomos and Narrative', *97 Harvard Law Review*, 4 (1983).

— 'Violence and the Word', *Yale Law Journal*, 95.8 (1986), 1601–29, https://doi.org/10.2307/796468

de Waal, Frans, *Are We Smart Enough to Know How Smart Animals Are?* (New York, NY: W. W. Norton, 2016).

Forthman, Debra L., Lisa F. Kane, David Hancocks, and Paul Waldau, eds, *An Elephant in the Room: The Science and Well-Being of Elephants in Captivity* (North Grafton, MA: Center for Animals and Public Policy, 2008).

Frankl, Viktor E., *Man's Search for Meaning: An Introduction to Logotherapy*, 4th edn (Boston, MA: Beacon Press, 1992).

Haskell, David, *The Songs of Trees: Stories from Nature's Great Connectors* (New York, NY: Penguin, 2017).

Kimmerer, Robin Wall, *Braiding Sweetgrass: Indigenous Wisdom, Scientific Knowledge, and the Teaching of Plants* (Minneapolis, MN: Milkweed, 2013).

Leopold, Aldo, *A Sand County Almanac, with Essays on Conservation from Round River* (New York, NY: Ballantine, 1991).

Olmert, M. D., *Made for Each Other: The Biology of the Human-Animal Bond* (Cambridge, MA: Da Capo Press, 2009).

Orr, David, *Earth in Mind: On Education, Environment, and the Human Prospect* (Washington, DC: Island Press, 1994).

Roszak, Theodore, 'On Academic Delinquency', in *The Dissenting Academy*, ed. by Theodore Roszak (New York, NY: Vintage, 1968), pp. 3–42.

Shubin, Neil, *Your Inner Fish: A Journey into the 3.5-Billion-Year History of the Human Body* (New York, NY: Pantheon Books, 2008).

Waldau, Paul, *Animal Studies — An Introduction* (Oxford: Oxford University Press, 2013).

SECTION II

THINKING IN
LATIN AMERICAN FORESTS

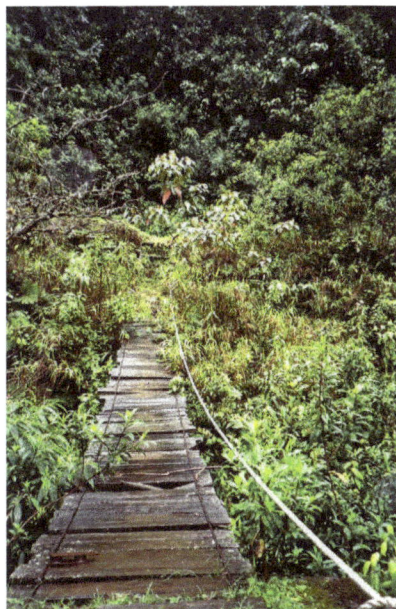

Fig. 4 Suspension Bridge in Cloud Forest Near Mindo, Ecuador. Photo by Ayacop (2007), Wikimedia, public domain, https://commons.wikimedia.org/wiki/File:Mindo-Cloud-Forest-15.jpg#/media/File:Mindo-Cloud-Forest-15.jpg

4. Anthropology as Cosmic Diplomacy

Toward an Ecological Ethics for Times of Environmental Fragmentation

Eduardo Kohn

Introduction

I'm an anthropologist. My job is to immerse myself ethnographically, to chart relations, and to find new ways to listen. Garbed in the flesh and skin I've come equipped with, protected by my words and the stories I weave together with them, I take these tools that make me human into the world we call 'the field'.

Perhaps today our vocation's name might feel a bit outdated given that our task to immerse ourselves can take us to fields where not all of the beings we encounter are of the anthropic sort. Working as I do in and around Indigenous communities of the Ecuadorian Amazon threatened by the destruction of ecologies — of relational worlds — these more-than-human beings include plants, animals, and even, and perhaps especially, spirits. Learning to listen to these other kinds of others has forced me to divest myself of some of the human trappings that equip me and to thus travel beyond the schemas through which I normally think.

Despite the fact that its theories are fashioned almost entirely from our human equipment, anthropology, thanks to its immersive method, is a vocation that can uniquely open us to the worlds these other kinds

 https://doi.org/10.11647/OBP.0186.04

of beings inhabit. Our attempts to grapple with what we learn there, as well as how we learn it, can allow us to capacitate other kinds of concepts, perhaps even, as Manari Ushigua, my Sapara colleague suggests, other kinds of gods. Giving life to these other kinds of concepts involves understanding thoughts from one world in terms of those from another with a view to grasping the emergent concepts that might unite these thoughts as one. In this sense a synonym for anthropologist is *yachak*, or 'knower', which is the Quichua word the humans I work with use for shaman.

Moving among worlds is not merely a scholarly endeavor. It is a political act. We do so in order to recognize the ways we take part in that larger flow of life that is today under grave threat. In this sense, another synonym for anthropologist might be what Bruno Latour calls a 'diplomat', more accurately, a *cosmic diplomat*; for the aim of moving among worlds is to find ways to avoid a cosmic — by which I mean an ecological — cataclysm.[1]

In recognition of the ways in which culture is now a force of nature, some geologists have proposed the term Anthropocene for the geological epoch in which we live. Living in the so-called 'time of humans' requires us to rethink what we mean by the human, and to rethink for the future (this epoch is far from over) a kind of ethics appropriate to a time in which separating humans from nonhumans is no longer practically or metaphysically conceivable. This involves recapturing the shamanic and diplomatic valences of the anthropological vocation — donning other kinds of clothing and equipping ourselves with other kinds of tools, not all of which are of the human sort. Working, as I do, in the Amazonian rainforest, my task as cosmic diplomat is to allow *sylvan selves* — the plants, animals, and especially spirits that also make their homes in the forest — a mode of expression that can be heard within our scholarly, biological, political, and legal idioms.

Thinking Forests

With this end in mind I wrote a book called *How Forests Think*, based on long-term fieldwork in and around Ávila, a Quichua-speaking

---

1   Bruno Latour, *An Inquiry into Modes of Existence* (Cambridge, MA: Harvard University Press, 2013) pp. 28–46.

Runa community in the northwestern part of Ecuador's Amazon region.[2] When I say that forests think, I don't mean it as a metaphor, nor am I referring to a culturally embedded belief. The claim is rather part of a diplomatic effort to convince you of the reality of things that can sometimes go unnoticed given the limits of certain metaphysical assumptions that form the axiomatic foundations for Western scholarly thought, including anthropological thought.

The underlying assertion in my statement that 'forests think', is that life is mind — that life is thought. What we share with other beings isn't so much our bodies, but our capacity to think. Mind here refers to that process, wherever in the universe it is found, of learning by experience. Evolutionary dynamics, in this sense, are mental dynamics because they imply the ways in which a lineage, over time, and via natural selection, learns something about its environment. Wings, as they evolved, have come to increasingly represent something about the currents of air on which they glide, for those lineages of organisms that have them. This is an example of thought; it is a kind of intelligence. One could say, in philosopher Charles Peirce's terms, that it is a 'scientific intelligence'.[3] This kind of thought, like all true — by which I mean living[4] — thought, does something. Flying becomes a new mode of being for a new kind of avian creature. When thought is alive it is because it makes this kind of worldly difference.

There are places in the world where this kind of mental dynamic is amplified — places where there is more mind, more thought, places that exhibit more scientific intelligence. One such place is Ecuador's 'megadiverse' Amazon region. If lives are minds, these dense tropical ecosystems would be sites for the emergence of ecologies made up of an unprecedented multitude of minds, thinking an equally unprecedented multitude of thoughts.

We humans have developed many techniques to amplify this kind of thinking. The great success of the scientific method is due, in part (I'm

---

2 Eduardo Kohn, *How Forests Think: Toward an Anthropology beyond the Human* (Berkeley, CA: University of California Press, 2013), https://doi.org/10.1525/california/9780520276109.001.0001

3 Charles S. Peirce, *Collected Papers of Charles Sanders Peirce* (Cambridge, MA: Harvard University Press, 1931–35), p. 2.227.

4 By saying that true thoughts are living thoughts I mean that for thought to be truly thought it must be alive, which, in terms of thought, means that it is constantly being re-interpreted by subsequent thoughts in ways that make a difference.

well aware of the power structures through which science operates), to the fact that it is a form of thinking that can self-consciously tap into the ways in which evolutionary dynamics themselves learn by experience. That is, the scientific method, and the emerging community that thinks through it, harnesses and amplifies the ways in which the world itself thinks. It is its own evolutionary dynamic that has learned to think by listening to the scientific intelligence already operant in the living world.

But this is not the only kind of science. Amazonian shamanistic practices that involve the ingestion of the psychedelic decoction *ayahuasca* or the cultivation and interpretation of dreams, to give two examples, are also sciences in the sense that they constitute specific techniques to accelerate and amplify a process of learning by experience. Their great advantage over other sciences is that their particular form of learning involves the systematic disruption of some of our human schemas for thinking. That these practices have unfolded in that place on our planet with the richest proliferation of nonhuman minds is no coincidence, and it makes them a privileged form of thinking scientifically with the scientific intelligence inherent to life. I find the etymology of the word psychedelic productive to think with. From the Greek *psychē* (soul or mind) and *dēloun* (to make manifest), *ayahuasca* makes manifest to us the mind of those thinking forests that are themselves, mind-manifesting.

So, forests think. But how do they think? The biggest obstacle we face in grasping this kind of thought is that we confuse what thinking is with a specifically human form of thinking that tends to erase other more expansive, but more fragile, forms of thought. What makes human thinking distinctive is a representational dynamic that, following Peirce, can be termed 'symbolic'. Symbols come to mean by virtue of the relationships they have to systems of other symbols, which form the interpretive contexts that gives them meaning. The English word 'dog', for example, refers to the animal in question indirectly thanks to a prior relation to the system of symbols that give it meaning. Thinking in symbols is what makes us so special as humans; it is the basis for language, culture, and consciousness.

But we are also open to other forms of thinking that reach well beyond the human, forms of thought that we share with all other living selves. This kind of thinking has another kind of dynamic whose logic is based more on the image than on the word. It traffics in two non-symbolic representational modalities, those that are 'iconic' and those that are 'indexical'. Of these, indices are the easiest to grasp. An index is a kind of sign that corresponds to or correlates with something it is not. For example, a monkey's cry of danger is not the potentially dangerous entity it indicates.

Indices, however, are the product of complex interactions among a much more counterintuitive iconic sign process that underlies it. Icons refer to their objects of reference, not by pointing to them — they don't actually in and of themselves refer at all, and they therefore exist at the very margins of semiosis and of thought — but by sharing in and of themselves something of the properties of the object in question. If ontology, in the classical sense in which I use the term, is the exploration of those realities that are independent of how we humans might relate to them, then iconicity, being the kind of sign that is what it is regardless of how it relates to its object, might confer an interesting vantage from which to explore such realities.

Indices and icons make up the form of thinking proper to forests. When, for example, a spot-winged antbird's alarm call *points to* a jaguar's presence, and a hunter simulates that call he heard in a way that *resembles* it, both partake in a form of thinking that is imagistic. And when we cultivate our dreams or take *ayahuasca* we are also thinking with and like forests, for these techniques temporarily break parts of the symbolic systems that house and sustain us as humans, permitting our thoughts to rejoin that kind of thinking that goes beyond the human. This form of thinking, which, as living selves, is something that is also ours, I call sylvan, wild, or savage, as in a *sauvage*. Sylvan thinking (a veritable *pensée sauvage*),[5] like all good scientific intelligences, amplifies, and thus makes available for further thought, certain properties of the sylvan worlds with which it thinks; it has a psychedelic potential.

To my mind, the phenomenon we are calling the Anthropocene is an actualization of the dualism inherent to symbolic thinking. Symbolic thought creates virtual and relatively closed thought-worlds that relate indirectly to the more concrete worlds to which they also refer. Agriculture, animal husbandry, the rise of cities and states, the industrial revolution, the accelerated flow of capital and information are

---

5 The reference is to the title of Claude Lévi-Strauss' classic, translated in English as *The Savage Mind* (Chicago, IL: University of Chicago Press, 1966).

increasing, perhaps historically contingent realizations of this human tendency to create realms of 'culture.' These realms are more and more separable — perhaps alienated — from 'nature', to such a degree that culture can eventually actually become a 'force' of nature.

A great danger of being human is to get too caught up in what makes us distinctively human. Donald Trump's particular brand of me-first 'thoughtlessness', which aligns individual, national, racial, gender, and even species narcissisms with in an ever-expanding arc exhibiting a brutal fractal-like symmetry, is a chilling consequence of this isolation from the worlds that hold us.[6] In this regard, the human sciences haven't helped. Conceptual tools that grow out of working with the distinctive symbolic properties of human thought (I'm thinking particularly of social construction in all of its variants) make it even more difficult to understand a way of thinking beyond the sort of dualism that pulls humans out of those worlds that both make us and are *not* us.

Harnessing the Logic of Sylvan Thinking

Given the ways in which our lives and thoughts are so entangled with dualism, how can thinking with forests help us? Sylvan thinking holds dualism in the sense that it is a form of thinking that is larger than the human. This can help us work conceptually with the connections we have to the nonhuman despite the separation that our distinctive forms of thought create. Cultivating sylvan thinking as an ethical orientation for the Anthropocene involves harnessing some of its more-than-human properties. I will briefly discuss four of them. Sylvan thinking involves: (i) images; (ii) absences; (iii) play; and, (iv) something I'll call 'generals'.

Sylvan thinking's imagistic qualities confer on it a host of counterintuitive properties. Consider, adapting an example from Terrence Deacon, the cryptically camouflaged Amazonian katydid *Cycloptera speculata*.[7] How did such a katydid come to look so much like a leaf? This does not depend on anyone noticing this resemblance — our usual understanding of how likeness works. Rather, its likeness is the product

---

6   Donna Haraway, *Staying with the Trouble: Making Kin in the Chthulucene* (Durham, NC: Duke University Press, 2016), https://doi.org/10.1215/9780822373780. I adopt the term 'thoughtlessness' from Donna Haraway (p. 36), following Hannah Arendt.
7   Terrence Deacon, *The Symbolic Species* (New York, NY: W. W. Norton and Company, 1997).

of the fact that the ancestors of its potential predators did *not* notice its ancestors. These potential predators failed to notice the differences between these ancestors and actual leaves. Over evolutionary time those lineages of katydids that were least noticed survived. Thanks to all the proto-cryptic katydids that were noticed — and eaten — because they differed from their environments, *Cycloptera speculata* came to be more like the world of leaves around it.

How this katydid came to be so invisible reveals important properties of iconicity. Iconicity, the most basic kind of sign process, is highly counterintuitive because it involves a dynamic in which two things are not distinguished. We tend to think of icons as signs that point to the similarities among things we know to be different. But semiosis does not begin with the recognition of any intrinsic similarity or difference. Rather, it begins with *not* noticing possible differences. It begins with indistinction or confusion.

Let me say something else about the imagistic logic that characterizes sylvan thinking: it is deeply personal. Icons share something in common with the objects they represent. In a way they *are* their objects. There is an emotional correlate to this — a feeling of identification, a feeling of knowing — a feeling of oneness. However, convincing others of this can be quite difficult. To 'get' an icon you have to feel it for yourself. In my lectures I often illustrate iconic thinking by having people guess at the meaning of a Quichua imagistic 'word', such as *tsupu*, which is used to describe an object making contact with and then submerging under water. I then contrast this word with other more standard conventional words in Quichua (which, being conventional, don't have this kind of sonic imagistic connection to what they mean). Once I tell them *tsupu*'s meaning, many people in an audience will immediately come to feel what it means. It is a likeness of an object plunging that they feel inside them. Invariably, however, some will not feel it, and no argument I can make will make them feel it. Sylvan thinking shares these qualities. The only way to grasp this imagistic logic is to feel it for yourself. Doing so requires a being/becoming sylvan, insofar as you need to find within you some of its qualities that you already share to iconically identify with its mode of being. This has important methodological implications for how we should go about thinking with forests, to which I will return.

Thinking with cryptic insects leads to my second observation about sylvan thinking: that it has an absential quality. We usually think of nature in terms of presence: matter, materiality, and existence are the foundations for our metaphysics. But absence is central to life; it is a kind of non-existence that is real.[8] Think of the ways in which such katydids are multiply absential. They have become 'invisible' (that is, absent) because they *re*-present (an absent) leafy environment. The environment is absent, in the sense that, after all, these katydids are not their environment. They are not, in fact, leaves. Katydids do this for (an absent) *future* generation — the future katydids in a lineage of katydids. They can do so thanks to the (absent) *dead* who were noticed and eaten by predators.

My third observation about sylvan thought is that it involves play. By play, I mean a dynamic in which previously tightly coupled means/ends relations are loosened such that something new can emerge. Play is ubiquitous in the living world. But this is because means/ends relations are intrinsic to the living world, and not just something we humans impose on it. In this technical Weberian sense, the forest is enchanted. By saying that life is semiotic, that forests think, I am also saying that function, representation, purpose, and telos — in short, ends — are part and parcel of the living world.

But if we think of means and ends as tightly coupled — transitive and deductive — there is no room for something new, for growth, for flourishing, which of course is also central to life. This is where play comes in. The biological production of variation is a form of play; Gregory Bateson's nip, that bite that denotes the bite but not that which the bite denotes (a ludic suspension of aggression he saw in dogs and other social mammals), is also a form of play; and any relaxation on selection creates a space for play.[9] Growth requires play in this sense. And we should remember that, for Claude Lévi-Strauss, the *pensée sauvage* is also a form of play in that it is a kind of thought that asks for no return.

The final observation about sylvan thinking is that it involves generality. Thanks to all the katydids that were not noticed there is now more 'leafiness' in this world. Not only are leaves that leafy but so too

8   Terrence Deacon, *Incomplete Nature: How Mind Emerged from Matter* (New York, NY: W. W. Norton and Company, 2011).
9   Gregory Bateson, *Steps to an Ecology of Mind* (Chicago, IL: University of Chicago Press, 2000) p. 180.

are some insects. Generality is a real property of the world — one that grows in the realm of life. Life proliferates generals. Through a process of constrained confusion living dynamics create *kinds*. Think of Jacob von Uexküll's tick, the one that is 'world poor' because it doesn't do a lot of differentiation.[10] By not discriminating between humans and deer, indiscriminately parasitizing both, confusing them, it creates a *kind* — the kind of being through which, for example, Lyme disease might pass. The world, then, is not just a continuum waiting to be categorized by human minds and cultures.

This logic extends to biological concepts such as the distinction between individual and lineage. It may be that only the individual exists, but the lineage is the reality that makes that existence possible. Any individual katydid is only what it is by virtue of a lineage that temporally exceeds it. This is true also of the species. It too has this kind of general reality. In this regard, the species is not unlike the Amerindian concept of the masters of animals. A master of animals is a being that is the protector and general instantiation of the species in question. All hunting passes through this generality. Hunters dream with or about this domain of the general in order to connect with the individual that will become meat. This generality is real even if its existence is only instantiated in the forest encounter.

The reality of forest spirits, then, is on par with the reality of a species or lineage. Out of an ecology of selves there emerges an ecology of spirits — or gods — as well. And this reality is not reducible to 'the social'. It is to this emergent spirit-life that we must also learn to attend. For these gods, or others like them, will be the ones who can *orient* us in the way that a kind orients an individual, and a dream orients the hunter. An ethical orientation for the Anthropocene must thus necessarily also involve a spiritual re-orientation. Spirits, gods, and souls are part and parcel of the sylvan thinking we need to inhabit once again.

The Politics of Sylvan Thinking

Having thought a great deal about sylvan thinking, and convinced that thinking with it can provide ways to think for our times, my current

---

10   Jakob von Uexküll, 'The Theory of Meaning', *Semiotica*, 42.1 (1982), 25–82, https://doi.org/10.1515/semi.1982.42.1.25

research projects focus on finding spaces of collaboration with others who seek to sustain and capacitate domains of sylvan thinking by tapping into their imagistic, absential, playful, and general logics.

This has brought me into close collaboration with a far-flung community of thinkers whose human members range from Indigenous leaders and shamans, to environmental activists, conceptual artists, and human rights lawyers. On the nonhuman side, it has led me to explore ways to think with the spirits of the forest, the obdurate animacy of *huaira* — wind — *alpa* — earth — as they make themselves present to me. This, in turn, has raised many questions: what methods should one develop to listen to these other beings? And, given that our modern metaphysical framework has relegated spirits to the realm of belief, how can one bring them back into concept-work and conversation without being branded a 'believer'?

I should say at the outset that Ecuador is a privileged place to cultivate an ethics of sylvan thinking for the Anthropocene. First off, as I've mentioned, this is because it houses an unprecedented amount of biodiversity, and diverse communities of people who continue to think with it, especially but not only, in its Amazonian forests, not all of which are, at least for the moment, in ruins. This kind of life and human forms of living with it are given unprecedented recognition in Ecuador's 2008 constitution, which was the first in the world to recognize the Rights of Nature. This constitution is also framed in terms of *Sumak Kawsay*, an idea of living well that is not based on the modern metrics of progress and unfettered economic growth, as well as a respect for Indigenous plurinationalism and self-determination.

As lofty as this document appears, its aspirations are rarely given a practical existence. Although written at the beginning of Rafael Correa's presidency, the Correa regime was characterized by an increasing suppression of alternative voices — sylvan and otherwise — and a ratcheting up of extractive policies and practices. Large-scale mining projects, roads, hydroelectric dams, and oil concessions have proliferated and many of these are funded by China, to whom Ecuador now has massive debts. Ecuador's 'neo-extractivist' tendency, as this logic is known in Latin America, runs counter to these innovative constitutional principles, as it has sought to feed a state whose top-down logic became increasingly amplified under the increasingly authoritarian Correa regime. If a vibrant democracy should resemble a dense forest, Ecuador is increasingly becoming a monocrop plantation. This is the terrain through which sylvan thought must learn to navigate.

As its own ethical practice, sylvan thinking would take the logic of the image as a legitimate form of knowing. It would cultivate absential dynamics as a kind of causal modality that is different from the exclusively 'push-and-pull' understanding of agency typical to our metaphysics. Its object would be to hold open the spaces of play from which it continuously emerges. And it would operate under the guidance of its own general emergent psychedelic properties, which, in other words, we might call *spirit*.

Bibliography

Bateson, Gregory, *Steps to an Ecology of Mind* (Chicago, IL: University of Chicago Press, 2000).

Deacon Terrence, *The Symbolic Species* (New York, NY: W. W. Norton and Company, 1997).

— *Incomplete Nature: How Mind Emerged from Matter* (New York, NY: W. W. Norton and Company, 2011).

Haraway, Donna, *Staying with the Trouble: Making Kin in the Chthulucene* (Durham, NC: Duke University Press, 2016), https://doi.org/10.1215/9780822373780

Kohn, Eduardo, *How Forests Think: Toward an Anthropology beyond the Human* (Berkeley, CA: University of California Press, 2013), https://doi.org/10.1525/california/9780520276109.001.0001

Latour, Bruno, *An Inquiry into Modes of Existence* (Cambridge, MA: Harvard University Press, 2013).

Lévi-Strauss, Claude, *The Savage Mind* (Chicago, IL: University of Chicago Press, 1966).

Peirce, Charles S., *Collected Papers of Charles Sanders Peirce* (Cambridge, MA: Harvard University Press, 1931–35).

von Uexküll, Jakob, 'The Theory of Meaning', *Semiotica*, 42.1 (1982), 25–82, https://doi.org/10.1515/semi.1982.42.1.25

5. Reanimating the World

Amazonian Shamanism

Frédérique Apffel-Marglin

Indigenous Amazonian peoples have been able to achieve results in the field of agriculture and healing that often surpass what modern humanity has been able to achieve.[1] Given the centrality of shamanism to Indigenous Amazonian people — with its ability to access knowledge through permeable, non-rational consciousness — we are encouraged to recognize that there are other modalities of cognition, in addition to the analytic and rational mind which focus with laser precision on some well-bounded aspect of reality. Analytic reasoning leads us to believe that the boundedness of the object of study is inherent to the object, rather than a result of our focusing on it or our observing it.

It is serendipitous, therefore, that the frontiers of western science are beginning to dovetail with the Indigenous worldview of sentience and meaning in nature. The belief in a 'reality out there', utterly distinct from human observers, is beginning to give way to a different modality.

As Karen Barad, quantum physicist and feminist philosopher declares: 'Meaning is not an ideality; meaning is material. And matter isn't what exists separately from meaning'. This statement asserts that meaning and matter are not two distinct, separate realities, the former

---

1   I mention only those two fields not because they are the only ones, but rather because they are the ones I am most familiar with. The architectural feats of Inca constructions are well known, and there are other such instances. See details about the pre-Columbian Amazonian anthropogenic soil known as Terra Preta do Indio in Brazil in chapter three of Robert Tindall, Frédérique Apffel-Marglin, and David Shearer, *Sacred Soil: Biochar and the Regeneration of the Earth* (Berkeley, CA: North Atlantic Books, 2017).

    https://doi.org/10.11647/OBP.0186.05

belonging to humans and their minds and the latter to an unconscious nonhuman world.[2]

In her ground-breaking book *Meeting the Universe Halfway*, Barad extends Niels Bohr's profound insight that the observed and measured object cannot be separated from the measuring and observing apparatus.[3] Barad shows that what we humans observe and measure is not an independent, external, given nature, but rather is what she calls an 'agential reality', namely an entanglement of observer and observed, of object and subject.[4] In this extension of Bohr's work, Barad shows that nature as existing completely separately from humans' observation of it simply 'disappears'.[5]

Barad's work on 'the disappearance of nature' is particularly powerful, since she has developed her theory of 'agential realism' based on her reading and extension of Bohr's understanding of the new quantum phenomena as recorded in his philosophy-physics papers.[6] Agential realism incorporates Bohr's fundamental insight that physical reality is a function of the agencies of observation, rather than pre-existing the measurements these agencies of observation produce. Or, as Michael Pollan, in reference to quantum physics, puts it, 'matter might not exist as such in the absence of a perceiving subject'.[7] This statement appears in his recent book on psychedelics. On that same page he also states the following:

> One of the gifts of psychedelics is the way they reanimate the world, as if they were distributing the blessings of consciousness more widely and

---

2   Karen Barad, 'Diffracting Diffraction: Cutting Together-Apart', *Parallax*, 20.3 (2014), 168–87, at 175, https://doi.org/10.1080/13534645.2014.927623

3   *Meeting the Universe Halfway* (Durham, NC: Duke University Press, Pantheon Books, 2007). Niels Bohr, *The Philosophical Writings of Niels Bohr, Vol. 3: Essays 1958–1962 on Atomic Physics and Human Knowledge* (Woodbrige, CT: Ox Bow Press, 1963), pp. 59–60.

4   Karen Barad, 'Reconceiving Scientific Literacy as Agential Reality, or Learning How to Intra-Act Responsibly within the World', in *Doing Science and Culture*, ed. by Roddey Reid and Sharon Traweek (New York, NY: Routledge, 2000), pp. 221–58, at p. 232.

5   Ibid.

6   In *Meeting the Universe Halfway*, Barad discusses in great detail the recently performed *gedanken* experiments, and offers a striking resolution of the measurement problem in quantum physics. For a fuller presentation of Barad's work, see chapter four in my book *Subversive Spiritualities* (New York, NY: Oxford University Press, 2011).

7   Michael Pollan, *How to Change your Mind: What the New Science of Psychedelics Teaches Us About Consciousness, Dying, Addiction, Depression, and Transcendence* (New York, NY: Penguin Press, 2018), p. 413.

evenly over the landscape, in the process breaking the human monopoly on subjectivity that we moderns take as a given. [...] Psychedelic consciousness overturns that view, by granting us a wider, more generous lens through which we can glimpse the subject-hood — the spirit! — of everything, animal, vegetal, even mineral, all of it somehow returning our gaze. Spirits it seems are everywhere. New rays of relation appear between us and all the world's Others.

Under expanded states of consciousness, humans are able to access knowledge otherwise beyond their reach. Shamans in the Upper Peruvian Amazon receive this knowledge directly from plants via the 'spirit' of these plants. When shamans are tasked with diagnosing and treating a malady, the plant shows them the cause of the malady, as well as what the treatment should be, specifying dosages and other protocols.

A similar reverence is shown in the practice of agroforestry, where one must prepare oneself to enter the forest by taking certain purges, including with psychotropic plants. As the leader or Apu of an Indigenous community with which my center collaborates stated to me:

[The forests] are our markets, our house, because there we find everything: medicines, food plants, animals, material for construction, and the forest is also a sacred space where we pray and ask permission and do rituals. One does not enter the forest just like that, one needs to take certain plants to purge and purify oneself so as to be in contact with the animals and be able to see the spirits.[8]

By taking psychedelic plants, Amazonian Indigenous people can receive knowledge from the various elements in the forest: trees, other plants, earth, fungi, animals. Indeed, they can receive knowledge from any element, be it water, air, fire, rocks, or the soil.

As did our pre-modern European ancestors, Amerindians understand the nonhuman world to be full of beings with consciousness and knowledge, and they therefore prepare themselves carefully to enter into communication with these beings. No activity is undertaken without consulting with the spirits of the place, of the forest. These communications are always respectful and tinged with an aura of sacrality, since the whole forest is a sacred place. Even a Westerner

---

8   Apu Lisardo Sangama Salas, cited by Tindall, Apffel-Marglin, and Shearer, *Sacred Soil*, p. 112.

such as myself, trained in modern universities, can receive extremely precise and clear knowledge while taking certain healing plants during shamanic forest retreats. I know from personal experience that the knowledge can be extremely precise, to the point, and transformative.[9]

We can confidently assert that similar ways of accessing information from what we moderns label 'the environment' is what produced the stupendous achievement of Terra Preta do Indio (Amazonian Dark Earths). Terra Preta do Indio is the Portuguese name for the anthropogenic pre-Columbian Amazonian soil re-discovered by archeologists in the last century, and declared by soil-scientists to be the most sustainable, fertile soil in existence, and one that can, in turn, sequester greenhouse gases from the atmosphere in the very substantial amount of 20%.[10] It is simply foolish to *a priori* disregard these methods of intra-action in the face of the ecological crisis initiated by modern society.[11]

The new quantum physics outlined by Barad opens a window onto the view that non-modern, non-Western peoples have been able to co-create ecosystems, soils, and medical healing — among other things — that in some ways distinctly surpass our modern Western approach of mastering or managing an insentient, mechanical nature. The view that these peoples did not achieve these things through an empirical process of trial and error, but rather through receiving information from the spirits, can now be understood as their ability to recognize that matter and meaning are always entangled. Furthermore, by labeling such sources of information 'spirit' or 'deity' or any other such non-empirical being, and by giving them names, it is acknowledged that we are of the same stuff as they, the stuff of matter and meaning. Such an understanding enables the possibility that we co-create the world, since we and it are of the same stuff.

The cosmologist Brian Swimme shares a revealing personal story about cosmological discoveries and his attempts to communicate them

9   I refer the reader to Jeremy Narby, *The Cosmic Serpent: DNA and the Origins of Knowledge* (New York, NY: Jeremy P. Tarcher/Putnam, 1999) as another source of evidence for the exact nature of shamans' knowledge, as well as for the impossibility of acquiring such knowledge through a process of trial and error.
10   See http://css.cornell.edu/faculty/lehmann/index.html
11   Karen Barad, 'Meeting the Universe Halfway: Realism and Social Constructivism without Contradiction', in *Feminism, Science, and the Philosophy of Science*, ed. by Lynn Hankinson Nelson and Jack Nelson (Boston: Kluwer Academic Publishers, 1996), pp. 161–94, at p. 188.

to others. When describing what he felts as he embarked on these new cosmological findings, people often asked Swimme if he used drugs. His initial reaction was to somewhat angrily reject this suggestion. This reaction was a response to what he perceived as the listener equating the feelings induced in him by cosmological discovery with the effects of hallucination induced by drugs. However, upon reflection, Swimme developed a rather different response. He concluded that alcohol and drugs are an intrinsic feature of consumerism, necessary for its sustainability. Consumerism is based on the basic assumption of the modern worldview, namely that the world is made of dead objects. These objects are, in Swimme's wry words, mostly 'unmanufactured consumer goods'.[12] The deliriously abundant glory of the natural world, of the cosmos, is reduced to an inert mechanism. Humans are of this world, created from and with it, and this western modern paradigm cuts us off from the extraordinary expressiveness of this living, sensuous, numinous world. We are left alone among our kind, bereft of this numinous and exuberantly varied part of ourselves. The nonhuman world, the cosmos, has agency, sentience, and more.

These and other such findings among several scientists have barely percolated within academia, let alone the wider society and culture. Most of the institutions of modern society are based on the old Classical Scientific paradigm, one that gives certainty and power over the nonhuman world to these institutions, and to those humans seen as being closer to nature. It is no surprise, therefore, that these institutions do not welcome with open arms the recent discoveries made by some scientists. The reaction is what the neo-Jungian psychologist James Hillman calls 'collective ego-defenses' of the repressed unconscious of modernity.[13] Modernity is still overwhelmingly in the grip of this dead world, which is, at the same time, a deadening, pathological world. Swimme suggests that 'hoping for a consumer society without drug abuse is as pointless as hoping for a car without axle grease'. In what follows he explains why:

> When humans find themselves surrounded by nothing but objects, the response is always one of loneliness [...] But isolation and alienation

---

12 Brian Swimme, *The Hidden Heart of the Cosmos: Humanity and the New Story* (New York, NY: Orbis Books, 1996), pp. 33–34.

13 See James Hillman, *The Thought of the Heart and the Soul of the World* (Dallas, TX: Spring Publications, Inc, 1992), p. 94.

are profoundly false states of mind. We were born out of the Earth
Community and its infinite creativity and delight and adventure. Our
natural genetic inheritance presents us with the possibility of forming
deeply bonded relationships throughout all ten million species of life
as well as throughout the nonliving components of the universe. Any
ultimate separation from this larger and enveloping community is
impossible, and any ideology that proposes that the universe is nothing
but a collection of pre-consumer items is going to be maintained only at
a terrible price.[14]

Today, these ten million species have been severely diminished, with the
largest extinction of species since the disappearance of the dinosaurs,
and one caused by humans, giving our geologic era the label of the
'Anthropocene'.[15] We are all in deep mourning, depressed and bereft
whether we are aware of it, or of its deep lying causes or not, or whether
we have repressed all of this to our collective unconscious.

For me, the imperative of our times is the need to heal ourselves, and
help heal our children and our ravaged earth. In other words, there is
urgent need for integral ecological healing, repairing the split between
nature and culture, between mind and body, and between mind and
heart. In this endeavor, we need to avoid the Charybdis of fundamentalist
rationality and materialism, and the Scylla of unquestioningly accepting
the received wisdom of parents, school, and community. We also need
to recognize that spirituality is at the very core of such an endeavor.

So, here, at the field campus of my non-profit organization
Sachamama Center for Biocultural Regeneration in the Peruvian
Upper Amazon where I am writing these lines, I have tried to create
a space where this split can be healed. Indigenous spirituality in the
Peruvian Upper Amazon is a union between ancestral shamanism
and Catholicism. It seems to me that with Pope Francis's ecological
encyclical *Laudato si'*,[16] as well as his apology to Indigenous peoples
of the Americas, the first steps have been made towards transforming
the horrific history of the brutal enforcement of Catholicism here into
a fecund union. With this encyclical, Catholicism has re-invested the
nonhuman world with not only intrinsic value but also with numinosity.

---

14 Swimme, *The Hidden Heart*, pp. 33–34.
15 On the 'Anthropocene', see also 'Anthropology as Cosmic Diplomacy: Toward an
     Ecological Ethics for Times of Environmental Fragmentation' by Eduardo Kohn in
     this volume.
16 Pope Francis, *Encyclical: Laudato Si'* (Rome: Vatican Press, 2015).

For the Indigenous peoples and mestizo curanderos (shamans of mixed Indigenous and European ancestry) here, Catholicism and ancestral Indigenous spirituality are not experienced as antagonistic nor as separated. This compatibility is in part enabled by this peoples' oral milieu, where the history of violent Catholic conversion has lost its horror through the gradual erosion of memory.

Bibliography

Apffel-Marglin, Frédérique, *Subversive Spiritualities: How Rituals Enact the World* (New York, NY: Oxford University Press, 2011).

Barad, Karen, 'Meeting the Universe Halfway: Realism and Social Constructivism without Contradiction', in *Feminism, Science, and the Philosophy of Science*, ed. by Lynn Hankinson Nelson and Jack Nelson (Boston: Kluwer Academic Publishers, 1996), pp. 161–94.

— 'Reconceiving Scientific Literacy as Agential Reality, or Learning How to Intra-Act Responsibly within the World', in *Doing Science and Culture*, ed. by Roddey Reid and Sharon Traweek (New York, NY: Routledge, 2000), pp. 221–58.

— *Meeting the Universe Halfway* (Durham, NC: Duke University Press, Pantheon Books, 2007).

— 'Diffracting Diffraction: Cutting Together-Apart', *Parallax*, 20.3 (2014), 168–87, https://doi.org/10.1080/13534645.2014.927623

Bohr, Niels, *The Philosophical Writings of Niels Bohr, Vol. 3: Essays 1958–1962 on Atomic Physics and Human Knowledge* (Woodbrige, CT: Ox Bow Press, 1963).

Hillman, James, *The Thought of the Heart & the Soul of the World* (Dallas, TX: Spring Publications, Inc, 1992).

Narby, Jeremy, *The Cosmic Serpent: DNA and the Origins of Knowledge* (New York, NY: Jeremy P. Tarcher/Putnam, 1999).

Pollan, Michael, *How to Change your Mind: What the New Science of Psychedelics Teaches Us About Consciousness, Dying, Addiction, Depression, and Transcendence* (New York, NY: Penguin Press, 2018).

Pope Francis, *Encyclical: Laudato Si'* (Rome: Vatican Press, 2015).

Swimme, Brian, *The Hidden Heart of the Cosmos: Humanity and the New Story* (New York, NY: Orbis Books, 1996).

Tindall, Robert, Frédérique Apffel-Marglin, and David Shearer, *Sacred Soil: Biochar and the Regeneration of the Earth* (Berkeley, CA: North Atlantic Books, 2017).

6. The Obligations of a Biologist and Eden No More

Thomas E. Lovejoy

This chapter is based on two previously published pieces, 'The Obligations of a Biologist', and 'Eden no more.'[1] 'Obligations' was written in 1989 (before social media and cell phones and luggage with wheels!) and the world was on an encouraging but unfulfilled trajectory of environmental consciousness and action. A lot happened in 1992 with the Earth Summit, the three conventions, the Global Environmental Facility (GEF) and, later, the Sustainable Development Goals. 'Eden no more' was written as the Convention on Biological Diversity published its first global biodiversity assessment (2019), highlighting the progress made in biodiversity and, to a much lesser extent, climate change. The situation was better than in 1989, or in 1980 when I did the first projection on species extinctions, but the challenges had grown substantially.

The Obligations of a Biologist

All I ever really wanted to do was explore the wonders of tropical nature, combining the advantages of a latter twentieth century perspective with the enthusiastic thrill of nineteenth century naturalist exploration. Yet, like many, including every contributor in this volume, I have been unable to ignore the havoc being wreaked upon the biology of our planet. In consequence, I turned my hand to conservation, and my science to conservation biology.

---

1   Thomas E. Lovejoy, 'The Obligations of a Biologist', *Conservation Biology*, 3.4 (1989), 329–30, https://doi.org/10.1111/j.1523-1739.1989.tb00235.x; 'Eden no more', *Science Advances*, 5.5 (2019), https://doi.org/10.1126/sciadv.aax7492, https://advances. sciencemag.org/content/5/5/eaax7492.full

https://doi.org/10.11647/OBP.0186.06

The global crisis will essentially be played out in the decade we are about to enter. The very intensity of the problems raises difficult tensions and complex questions about the proper role of science. The venerable British Ecological Society, or at least some of its leaders, takes the view that as a scientific society, it should have nothing to do with conservation. For example, in a letter in the February 1989 *British Ecological Society Bulletin*, L. R. Taylor and J. M. Elliott, honorary editors of the *Journal of Animal Ecology*, wrote: 'The British Ecological Society is a Scientific Society, not an Environmental Protection Society. [...] We are supposed to produce the factual information used for whatever purpose, including environmental protection, but equally for environmental destruction if that is where mankind is heading.'[2] Nevertheless, statements we would have made purely on a scientific basis in the past take on a policy significance in today's world. An awareness of this public role, whether sought by ourselves or thrust upon us uninvited, is essential. We do not help either science or society by evading our social responsibilities as experts.

It is the very basis of science that it progresses in a dialectical fashion: evidence, counterevidence, new interpretations, new facts, and testing of long-held points of view. Naturally, this can be a source of confusion. The general public, for example, must sort out, on the one hand, that the greenhouse effect is something that must be taken with utmost seriousness, and, on the other hand, that some disagreement persists as to whether the warming has begun and that it is impossible to state with any authority how climate will change in any particular spot. Because the human tendency for denial is so great, it is critical that scientists involved in environmental issues, as we are, put in true perspective the new developments and minority opinions that contradict generally held conservation beliefs.

A good example is set for us by those working on the greenhouse effect problem. When most scientists state that the chance of a catastrophic global climatic change is greater than 50%, they explain that they are making a professional judgment. It is proper to go on to explain, however, that nobody would board an aircraft judged to have such a likelihood of major problems and would be reluctant even if the airplane had only a 5% chance of failure.

---

2   L. R. Taylor and J. M. Elliott, 'Letter to the Editor: Animal Experimentation and Environmental Issues', *Bulletin of the British Ecological Society*, 20 (1989), 15–27, at 20.

I would further assert that science must take on an advocacy role with respect to environment. If science does not, we deserve and can expect the future censure of society, for indeed it is our responsibility, as those who understand best what is happening and what alternatives exist, to sound the tocsin about environmental deterioration and conservation problems in all their variety. As conservation biologists, we have a very special role, for the biota is the ultimate assay of the environmental health of our planet. In essence, we should always be the first to know there is a problem. And we need to build a margin of error into our recommendations. How can we expect to be aware of all there ever will be to know when we recommend a minimum population size for some endangered species? Should we not double the figure to hedge against the limitations of our current ignorance? If we explain what we are doing, we in no way compromise our scientific credibility.

What makes this so particularly difficult is that nothing in our training as scientists has prepared us for a perspective generated by the era of planetary environmental crisis. Because we are both advocates and scientists, we must fight to protect our capacity for searching self-criticism; this is the only way our science can remain true to itself.

Eden no More

The first official report of the Intergovernmental Science-Policy Platform on Biodiversity and Ecosystem Services (IPBES), released on 6 May 2019 in Paris, provides the first modern authoritative assessment of planetary biodiversity and related contributions of nature to people (CNP) — dubbed ecosystem services.[3] Ecosystem services are those charities of nature, both nebulous and tangible, that serve as the backbone of human well-being: food, fresh water, clean air, wood, fiber, genetic resources, and medicine.

The IPBES is being called the IPCC of Biodiversity, with the IPCC referring to the Intergovernmental Panel on Climate Change, the recognized assembly of the United Nations created in 1988 to provide global leaders with regular scientific assessment of the implications and risks of climate change. The IPBES, founded in 2012, came slow on

---

3   IPBES, *Global Assessment on Biodiversity and Ecosystem Services*, 2019, https://ipbes. net/global-assessment-report-biodiversity-ecosystem-services

the heels of the IPCC for a variety of reasons but in large part because grappling with, gathering data for, and analyzing the myriad features of global biodiversity and ecosystem services are astoundingly complex endeavors.

A scientific assessment of the state of biodiversity and ecosystem services in the context of climate reveals that all are inextricably intertwined, united yet dispersed, invaluable yet monetizable, reflecting nature in its holistic role as the bedrock of human civilization. The 2005 Millennium Ecosystem Assessment served as an early appraisal of the state of life on Earth.[4] The IPBES synthesis is today's report card, and it tells a short story: Eden is gone. While the planetary garden still exists, it is in deep disrepair, frayed and fragmented almost beyond recognition.

Not unexpectedly, the specific findings are depressing. More species are threatened with extinction than any time in human history. Ever growing human populations and their activities have severely altered 75% of the terrestrial environment, 40% of the marine environment, and 50% of streams and rivers. The health of freshwater biodiversity has been particularly neglected because freshwater is widely understood and managed more as a physical resource vital to survival rather than as the special and delicate habitat that it provides for an extraordinary array of organisms.

The primary drivers of negative trends are also no surprise: in descending order, these adverse impacts include rapid changes in land and sea use, direct exploitation of natural resources, climate change, pollution, and invasive species. Of monumental note is that, collectively, significant destructive forces arise from the actions of impoverished peoples living at the edges of society, working to eke out an existence often with little choice but to have minimal concern for environmental impact.

The role of climate change in biodiversity loss is also severely underestimated because of the lag between rising levels of CO_2 concentration and the equivalent accumulation of the radiant heat that leads to warming and biological impact. Ironically, climate change is also, in part, the consequence of biodiversity destruction: the amount

---

4   Millennium Ecosystem Assessment, *Ecosystems and Human Well-Being: Synthesis* (Washington, DC: Island Press, 2005), https://www.millenniumassessment.org/documents/document.356.aspx.pdf

of carbon in the atmosphere from degraded and destroyed ecosystems is now equal to what remains in extant ecosystems. The additional CO_2 emanating from the combustion of fossil fuels is in fact ancient solar energy that was trapped and converted by ancient ecosystems and is now being released in a geological instant.

While the IPCC reports have documented climate change and sounded warnings, the IPBES report highlights aspects of the degradation of planetary natural systems that equally warrant immediate attention and action. As dire as the findings in the assessment may be, they likely also hold the ingredients for possible solutions. For example, economists and decision makers are largely unaware of (or choose to ignore) the contributions of natural resources to the Gross Domestic Product (GDP) of Indigenous peoples or the poor; at the same time, many of those people are often equally reluctant to embrace the monetary value of local ecosystem services. Contributions of Nature to People (CNP) and ecosystem services are essentially two congruent valuation systems, and both are recognized by the IPBES assessment. The danger is that decision makers are often distant from the actual sites of valued biodiversity and ecosystems; as a result, they do not see actual monetized benefits from the sustainable use of natural resources, and so peg the value of these resources at, or near, zero.

Adding to the flaws in the calculus of conservation and sustainability is the surprising inattention to the value of new discoveries from biodiversity and ecosystems to life sciences research. For example, researchers recently discovered that a soil fungus in Nova Scotia can functionally disarm antibiotic-resistant bacteria, a discovery that could transform practices in medicine, agriculture, and beyond. About 70% of drugs used for cancer are natural or bioinspired products. The polymerase chain reaction aided by an enzyme from a Yellowstone hot spring bacterium may have generated close to a trillion dollars of benefit through rapid multiplication of genetic material. The list of treasures uncovered in the elements and processes of the natural world grows daily; at present, however, these kinds of contributions from natural resources to human health and life sciences are neither recognized nor accounted for and so are treated as free and without value.

The IPBES report findings are more than sobering: 35/44 assessed targets of the Sustainable Development Goals depend on authentic transformational change to reverse trends of degradation. The assessment concludes that the current course of planetary degradation can be altered only with preemptive and precautionary actions, strengthened laws and related enforcement, dramatic changes in economic and social incentives, increased monitoring of biodiversity and ecosystems, and integrated decision-making across sectors and jurisdictions.

These dramatic changes will need to be supported by leaders, who themselves must promote new ways of understanding the meaning of 'quality of life', ones that value consuming less, wasting less, conserving more, and engaging truly novel approaches to global resource conservation and management. New tools will need to include technologies, creative economic models, and future-facing patterns of social behavior that are respectful of the diversity of needs, cultures, and local resources across the planet. These tools will need to be designed and applied to manage land use, agricultural development, and resource distribution in ways that will feed everyone adequately without further destroying nature.

Happily, the publication of the IPBES assessment coincides with new and hopeful visions emerging from the conservation community that adjust the scale and impact of collective efforts upward dramatically. The Edward O. Wilson Biodiversity Foundation's goal of Half-Earth was one of the first, with the aim of conserving half of the planet's lands and seas to safeguard the bulk of biodiversity, including humans. The National Geographic Society has a goal to place 30% of the planet in protected areas by 2030. The Global Deal for Nature,[5] is essentially coincident with the One Earth vision from the Leonardo DeCaprio Foundation.

The story of the unraveling of the planetary web of life has been told for decades, well before Rachel Carson's prediction of silent springs.[6] With its publication, the IPBES assessment, however imperfect, is now the most complete and comprehensive synthesis to date on the state of the health of the planet with all its natural resources and potential for contributing to human well-being. Readers at all levels of government, in the for-profit sector, and in civil society should heed its warnings and act on its vision and recommendations in haste. Together, we now sit at

---

5   Eric Dinerstein et al., 'A Global Deal for Nature: Guiding Principles, Milestones, and Targets', *Science Advances*, 5.4 (2019), eaaw2869, https://doi.org/10.1126/sciadv. aaw2869

6   Rachel Carson, *Silent Spring* (London: Hamish Hamilton, 1963).

the fail-safe point and must decide what to do; collectively, all sectors must embrace the challenges raised by the assessment, rise to action, and do what we must do to ensure a viable future for our living planet and for humans and the extraordinary variety of life with which it and we are blessed.

Bibliography

Carson, Rachel, *Silent Spring* (London: Hamish Hamilton, 1963).

Dinerstein, Eric, et al., 'A Global Deal for Nature: Guiding Principles, Milestones, and Targets', *Science Advances*, 5.4 (2019), eaaw2869, https://doi.org/10.1126/sciadv.aaw2869

Lovejoy, Thomas E., 'The Obligations of a Biologist', *Conservation Biology*, 3.4 (1989), 329–30, https://doi.org/10.1111/j.1523-1739.1989.tb00235.x

— 'Eden no more', *Science Advances*, 5.5 (2019), https://doi.org/10.1126/sciadv.aax7492, https://advances.sciencemag.org/content/5/5/eaax7492.full

IPBES, *Global Assessment on Biodiversity and Ecosystem Services*, 2019, https://ipbes.net/global-assessment-report-biodiversity-ecosystem-services

Millennium Ecosystem Assessment, *Ecosystems and Human Well-Being: Synthesis* (Washington, DC: Island Press, 2005), https://www.millenniumassessment.org/documents/document.356.aspx.pdf

Taylor, L. R., and J. M. Elliott, 'Letters to the Editor: Animal Experimentation and Environmental Issues', *Bulletin of the British Ecological Society*, 20 (1989), 15–27.

SECTION III

PRACTICES FROM CONTEMPORARY ASIAN TRADITIONS AND ECOLOGY

Fig. 5 Yamuna River, near the Himalayas. Photo by Alexey Komarov (2015), Wikimedia, CC BY 3.0, https://commons.wikimedia.org/wiki/File: Yamuna_River_-_panoramio_(2).jpg#/media/File:Yamuna_River_-_panoramio_(2).jpg

7. Fluid Histories

Oceans as Metaphor and the Nature of History

Prasenjit Duara

Time and tide wait for no man.

The field of pre-history is said to be characterized by the absence of a writing system. History is the study of human activity over the last 5,000 years or so based on texts and artefacts. Indeed, it is only recently that non-artefactual recording processes, such as carbon dating, tree-rings, the age of dental remains and bones have begun to be used in historical studies. John McNeill has proposed the idea of 'superhistory', distinguished from the academic discipline of Big History, which locates human history within the larger story of evolution. He explains that 'Big History changes everything and changes nothing. It tells us that our species' story conforms to a larger pattern. But it does not change our species story.'[1] Rather, superhistory is the effort to link the history of humans to natural factors, beyond the specialized field of environmental history, and beyond the textual and artefactual. It will be a revolution in methodology, geared towards grasping how humans have interacted with the climate, and how the climate has interacted with humans. Mine is a related effort: to grasp the relation of historical time to natural time at a metahistorical level.

---

1 John McNeill, 'Historians, Superhistory, and Climate Change', in *Methods in World History: A Critical Approach*, ed. by Janken Myrdal, Arne Jarrick, and Maria Wallenberg Bondesson (Lund: Nordic Academic Press, 2016), pp. 19–43, at 22, https://doi.org/10.21525/kriterium.2.b

       https://doi.org/10.11647/OBP.0186.07

I seek to understand the temporality of historical flows through the model of oceanic circulations of water. I will seek:

1. to probe what I have called 'circulatory histories' as fundamental historical processes that are not tunneled, channeled or directed by national, civilizational or even societal boundaries, but are circulatory and global, much like oceanic currents. Processes emerging in one form in place A flow to many places, B, C, etc., where they interact with other local and trans-local forces to re-emerge often in place A, though recognized as something else.[2] Older cosmologies, rhythms, and technologies of organizing time were closely in sync with natural patterns, even though they were not reducible to the latter.

2. to de-couple historiographical time from historical time, and re-link the latter to the temporal medium of natural flows. The flow of historical time can be viewed not only through the metaphor of ocean flows, but is also naturally continuous with the latter. The basic medium of temporal contact and intertemporal communication in historical time are the natural (including organic) and built environment which serve not only as its infrastructure but also its provisioner. Historical time is co-created by actors and actants, much as the ocean waters are the medium and co-creators of its flows with oceanic beings — both organic and inorganic.

3. by disclosing the metaphorical and natural links between historical and oceanic time, I want to reveal the consequences for the world as we know it, of the growing gap between historiographical time on the one hand, and historical and natural time, on the other.

---

2   In addition to my own *The Crisis of Global Modernity: Asian Traditions and a Sustainable Future* (New York, NY: Cambridge University Press, 2015), see chapter 2, https://doi.org/10.1017/cbo9781139998222, from which I cite some of my examples, see also Jack Goody, *The Theft of History* (New York, NY: Cambridge University Press, 2006), https://doi.org/10.1017/cbo9780511819841; and Kapil Raj, 'Beyond Postcolonialism... and Postpositivism: Circulation and the Global History of Science', *Isis*, 104.2 (2013), 337–47, https://doi.org/10.1086/670951

Circulatory Histories and Oceanic Flows

In my recent work, I have introduced the idea of 'circulatory history' — a kind of history that is interested principally in the flow of time.[3] Circulatory, in my usage does not necessarily reference a return to a starting point, but a movement or distribution from place to place. Historians have engaged with various conceptions of time, including the phenomenological, whereby different societies experience time differently. National and civilizational histories engage with evolution within a teleological framework of progress. The advocates of Big History and *Journey of the Universe* conceive historical time as embedded in processes of evolutionary complexity. But let us begin by asking what media — bodies, vehicles, and agencies — best allow us to recognize and measure the flow of time. The first candidate would be sunlight, with its diurnal and seasonal cycles. Another natural candidate is water, and may be more interesting for historical time, because, although water is a re-cycling element, we never step into the same water twice.

My interest in circulatory histories coincided with the burgeoning of inter-Asian studies, including important work by historians of South Asia who wrote about connected and entangled histories. Overwhelmingly, the picture we get of Afro-Eurasia is of a deeply interconnected historical sphere populated by sprawling networks, expanding and contracting empires, traveling ideas and practices, circulating microbes and species of all kinds. Our aim is to take stock of what kinds of intellectual, conceptual, and perhaps even epistemological significance this genre of trans-border work can have, not simply for Asian studies, but globally.

Since the publication of *Rescuing History from the Nation* in 1995, I have sought to dislodge historical writing from serving as the instrument of the nation's sovereign legitimacy.[4] The rationale for this is both simple and deep: the nation-form has been the dominant mode of identity for most of the world over the last couple of centuries, and it is structured to engage in a competitive race for global resource domination; in turn, it has led most visibly to two World Wars and to the ravaging of the global environment as we enter the Anthropocene.

---

3   Duara, *The Crisis of Global Modernity*.
4   Prasenjit Duara, *Rescuing History from the Nation: Questioning Narratives of Modern China* (Chicago, IL: University of Chicago Press, 1995), https://doi.org/10.7208/ chicago/9780226167237.001.0001

The forces for global cooperation and checks against predatory activities upon people and nature have been much weaker, in great part because of the nationalist imperative for GDP growth and the assemblage of interests legitimated by this imperative. In turn, I have argued that national histories are the principal means of establishing the imagined solidarity and destiny of the nation. I try to show here that history is by no means linear and tunneled, predestined to tell the story of the nation.

The oceanic metaphor of historical time allows us to grasp how historical ownership of science, technology, culture, civilization — the question of sovereignty itself — can only be sustained when historical process or flow is separated from historiographical understanding (i.e., not only in the historical discipline). When we attend to temporalities of different processes, we recognize that human history shares significantly with other organic and natural processes and that it is a collective planetary heritage. This recognition is imperative if we are to address the problems of planetary sustainability.

This, however, is not to say that there are no subjects — or agency, now recognized to be distributed between different humans and nonhumans — in history that seek to control or shape processes, but, beyond a point, the process escapes these subjects. Analytically, we need to distinguish between *historical* time and *historiographical* time, which includes purposive reflexivity. To be sure, the two are practically difficult to distinguish because this reflexivity also shapes the process. Nonetheless, the process exceeds the determinate purpose of the actor as it emerges in the circulatory flow of historical time. Historical time flows on, shaping and being shaped, carrying with it, to paraphrase the process philosopher Alfred North Whitehead, the 'many' from the 'disjunctive universe' which it gathers.[5] And in its carrying, there are also memories and brandings that are cognized by some and recognized

---

5   Methodologically, I follow *process philosophy*, particularly the idea of 'emergence' as delineated in the work of Alfred North Whitehead. Emergence is a form of creativity — which, for Whitehead, is the ultimate principle (also known as God). He writes that '"Creativity" is the principle of *novelty*. An actual *occasion* is a novel entity diverse from any entity in the "many" which it unifies. Thus "creativity" introduces novelty into the content of the many, which are the universe disjunctively. The "creative advance" is the application of this ultimate principle of creativity to each novel situation which it originates' (my emphases). Thus, historical processes and events may be seen fundamentally as emergences. Alfred North Whitehead,

by others as a return. But for a start, we need to analytically differentiate the reflexive moments from the process itself.

Historical time is not fundamentally different from the flow of time in nature, which too remains irreversible.[6] The flow of historical time is expressed in routine repetitive acts (never exactly identical) as well as the gathering or morphing into events caused by global interactions and contingencies, human and natural.[7] The model of natural processes that I find most useful to understand history is the circulatory flow of oceanic *water*. Unlike rivers, they are not tunneled and bounded; their channeling is more interactive.

Ocean currents develop in interaction with changing atmospheric conditions of heat and wind, geological features and tidal activity. The Coriolis Effect, trade winds, the Gulf Stream, Equatorial Currents and Counter Currents, El Niño, La Niña, monsoons, cyclones, tsunamis, upwellings, and thermohaline circulation are some of the well-recognized oceanic processes. The oceans and seas are realms where spaces and temporal processes interact at varying scales. Take, for instance, the Mediterranean Sea — a waterbody that has been well-studied due to the fact that it is relatively enclosed. This Sea is a microcosm of an ocean and like it, has surface, intermediate and deep-water masses, the circulatory patterns of which are relatively autonomous, but also influence the North Atlantic circulation regime. The geography of islands and their shelves affect the circulation of these waters significantly. Thus, the converging of the Sardinian and Tunisian shelves directs the inflowing Atlantic waters southwards whereas decaying eddies in the north are constrained to flow northwards off Western Corsica.[8]

Process and Reality, ed. by David Ray Griffin and Donald W. Sherburne (New York, NY: Free Press, 1985), p. 21.

6   In the Newtonian conception of time, natural processes can be reversible, but the temporal medium in which they occur is *absolute time*, which is not reversible; it depends on nothing external and moves in a linear way. See Brent D. Slif, *Time and Psychological Explanation* (Albany, NY: SUNY Press, 1993), pp. 273–74. The phenomenological conception, while not denying the idea of absolute time, finds it inadequate to the task of explaining how consciousness experiences a temporal object.

7   Routine repetitive acts — for instance, the activity of institutions — to be sure are still emergent occasions, because they are separated by 'degrees of difference' not necessarily visible in everyday activities.

8   T. M. El-Geziry and I. G. Bryden, 'The Circulation Pattern in the Mediterranean Sea: Issues for Modeller Consideration', *Journal of Operational Oceanography*, 3.2 (2010), 39–46, https://doi.org/10.1080/1755876x.2010.11020116

Oceans reveal circulatory currents of differing temporalities and effects, dependent on the diverse conditions they travel through. Surface currents are faster moving, because they carry heat and are shaped by winds; eddies are still faster, and more temporary gyres churn up smaller spaces. Deep currents are heavier, because water becomes colder at the poles and is pulled down by salinity and gravity. Nonetheless, deep currents also flow across the various oceans and cycle through roughly once every thousand years.

Over the last several decades, this deep-water flow across the oceans has been understood as a 'conveyor belt', reflecting the relatively stable pathway through which the warm waters pushed to the poles, overturned and made its way across the oceanic world. Recently, scientists have 'deconstructed' the model of the conveyer belt, suggesting that while overturning and coursing remain true, the conveyor belt idea was too simple. Rather, several different pathways have appeared, formed by surface eddies and wind-fields.[9] This suggests that their temporality — the rate of flow and the types of activity they produce — has been maintained at the level, but they are also interactive with other geo-atmospheric forces. Even the deepest level is affected, and, in turn, affects the rest. The impact of anthropogenic activity on the conveyer belt has alarmed the scientific community in recent years. The massive polar ice melt has already and, is expected to, increasingly lighten and warm the polar water with excessive freshwater and thus to slow (or even halt) the flow of the conveyer belt carrying tropical waters to the north. One major consequence of this is the generation of colder weather in North America and Europe.

Compare currents to historical processes — ideas, practices, and material — that flow through time and space. Let me cite a case of a circulatory ideational complex across continents over the last two hundred years. In 1833, Raja Ram Mohan Roy, a polyglot thinker and reformer, deist, and Unitarian, who is often called the 'father of modern India', was visiting Bristol in the United Kingdom. In Salem, Massachusetts, at the time, Unitarians were circulating a locket with a curl of his hair in preparation for his visit, which, however, never

---

9   M. Susan Lozier, 'Deconstructing the Conveyor Belt', *Science*, 328.5985 (2010), 1507–11, https://doi.org/10.1126/science.1189250, http://science.sciencemag.org/content/328/5985/1507

happened, since Raja Ram Mohan died in Bristol that year. New England Transcendentalists, particularly, Henry David Thoreau and Ralph Waldo Emerson, read Roy's translations of the Upanishads and the principal Vedas, texts they deeply admired and cited profusely. American transcendentalists influenced a wide range of global ideas and practices, including abolitionism, proto-environmentalism, and civil disobedience founded upon transcendentalist conceptions of self-cultivation of the powers of the mind and consciousness of ultimate reality. Thoreau's *Civil Disobedience* (1849)[10] influenced many people, including Leo Tolstoy, who in turn was an important influence on Mahatma Gandhi. In the 1890s in South Africa, Gandhi adopted the phrase 'civil disobedience' as the English version of his *satyagraha* (truth force) experiment. During Martin Luther King's civil rights movement in post-war USA, it was Gandhi and not Thoreau who was seen as its patron saint.

We can continue to trace this circulatory current, which merges, re-emerges, submerges, converges, and de-merges with various related or novel processes right up into the present. Thoreau's insights were carried on by spiritual naturalists — John Muir (Sierra Club), Aldo Leopold, and Arne Naess — and these insights today have transformed and emerged as a significant American environment movement (although with many different channels). E. F. Schumacher, Gary Snyder, the deep ecologists, and feminist ecologists, among others, have been influenced by Asian and Indigenous traditions. Many of these ideas around environmental spiritualism and moral protest have cross-fertilized with movements of Indigenous people, forest dwellers, civil society and religious groups across the world, climaxing (at least for millions of Catholic schoolchildren) with Pope Francis's radical encyclical of 2015 on ecology and justice.[11] What was for almost two centuries a sub-cultural and inconspicuous 'countercurrent' may yet swell into a movement of significance.

On a different level, consider the various temporal scales of historical processes. *Modern nationalism* (which developed symbiotically, for the

---

10 First published under the title 'Resistance to Civil Government', in *Aesthetic Papers*, ed. by Elizabeth Peabody (New York, NY: G. P. Putnam, 1849), pp. 189–213.

11 See Mary Evelyn Tucker and John Grim, 'Four Commentaries on the Pope's Message on Climate Change and Income Inequality', *The Quarterly Review of Biology*, 91.3 (2016), 261–70.

most part, with competitive capitalism) has developed as the axiomatic principle of legitimacy globally over the last two hundred years. The nation-form built around the *self-other binary* is the most enduring circulatory feature that has permeated all parts of the world which emerged from empires and other political forms built around more complex forms of belonging. The ideal nation-form is a confessional form that compacts *people-state-culture* for competitive control of global resources. Its immediate predecessor was the confessional state of the Reformation and Counter-Reformation in Europe, where a compact of *church-state-believer* believed itself to be saved as the chosen people, and other(s) to be damned.[12]

The ecology that sustains this doxic and *durable temporality* from the early nineteenth century has to do with the *system* of nation states that has been its most necessary condition for over two centuries. The fundamental raison d'être for the nation state is competition, even if competition alone is insufficient to account for nationalism in a particular time and place. At this level, the identitarian polity that is the nation, is *mediated* by a host of other forces such as religion, language, political regime, historical relations, etc. — forces that are often changing and mixing. Note that the institutions of the capitalist competitive order have not always been the most durable formations; consider, for instance, the period of Soviet and Maoist socialism. I believe that Maoism itself needs to be understood within a world-order of competitive states that ultimately pushed China towards capitalism.

Emergent historical forces or currents shaping societies at this level of mediation possess a kind of *middling temporality*. The temporality may be seen in the *mediatory* form of Chinese nationalism which changed in accordance with the change in China's place in the international order during the 1980s: simply put, a change from a Maoist socialist

---

12 The making of modern nation states since the nineteenth century is a complex and multidimensional process. However, two aspects, in particular, are important here. Firstly, nation states and nationalists seek to create a centralized, homogenized, and mobilizable political body to gain advantage in the competition of global resources for domestic growth. Secondly, nations are themselves products of circulatory forces morphologically similar to each other generated by importing and exporting 'best practices', to use an anachronistic term. See Prasenjit Duara, *Sovereignty and Authenticity: Manchukuo and the East Asian Modern* (Lanham, MD: Rowman and Littlefield, 2003).

state, to a globally participating market society. It changed gradually from the socialist model of the civic nation state that was built, however rhetorically, upon the fraternity of nationalities within and socialist and third-world internationalism abroad, to an ethnic model of privileging the culture of the Han majority. In practice, this shift was also facilitated by the need to attract powerful overseas Chinese capitalist networks based on Chinese culturalism and Confucianism. At the same time, the relative weakness of development in the western regions of China and among the ethnically marginal communities also fostered ethnic nationalism among these minorities that we are witnessing daily today. Finally, at the most *variable* level, like eddies and gyres, nationalism can function as an ideology, as political strategy, as mobilization politics and as ideals and dreams, changing according to contexts and constituencies.

The Nature of History

What we call history is fundamentally a natural process upon which humans have created artificial technologies of recording, including, of course, the exclusive prerogative of historical writing and reflection. What can we learn from re-embedding history in the natural process?

Natural processes too register their activities, whether in geological layers, in tree-rings, in DNA (which is a record of our species' epidemiological history), and, not least, in memory, language, and practices — the record of our social history. These are natural or elemental media for recording temporal processes which the inscriptions of historical humans mimic. Beings in time, of course, cease to exist; but not without registering their presence or trace — whether for functional purposes or not. To be sure, traces and records must be recognized and recalled. Biological organisms also leave information for the species and for others.

Human reflexivity and the technologies it has generated are said to distinguish historical knowledge from data produced by other organisms. But the sophisticated technologies of scientific observation reveal that the nonhuman world is constantly registering and responding to environmental changes, faster than ever in the Anthropocene. The

Star Moss Cam is a sensor technology that does not merely sense mosses over time, but observes how the moss itself is a sensor that is detecting and responding to changes in the environment.[13]

Thus, Susan Schuppli argues, matter itself bears witness to events. It registers and documents change internally within its substrates and molecular arrangements, and is often expressed as a material aesthetic, increasingly in 'dirty pictures', be it of black carbon on polar snow, radioactive leakage or oil spills that transform the Arctic. Inuit hunters who followed a predictable ecological sign system have lost their moorings and capacity to hunt not only with the rapid changes in the snowscape, but because the geo-atmospheric changes have brought about new optical regimes: the sun appears to be setting farther west and the stars seem to be no longer where they were.[14] The naked eye trained for generations to scope for life, can no longer do its job.

While many organisms may not have reflexive capacities, they certainly have cognition and communication. Just as the terrestrial earth, the ocean too is a storehouse of records and information for species and interactive organic forms. Cetaceans or marine mammals like dolphins and whales appear to differentiate sounds — phonations in water travel four times faster than in air — possibly identifying who must be responding to whom through the rapidly intensifying — and life destroying — din in the ocean.[15] A recent report on whale songs reveals that whales do not communicate only for mating purposes, but are constantly transforming and evolving forms of communication across hundreds of kilometres.[16] Some even claim that dolphins have an aquatic public sphere! In the words of John Durham Peters:

---

13   For more on the Star Moss Cam, see Jennifer Gabrys, 'From Moss Cam to Spillcam: Techno-Geographies of Experience', in *Program Earth: Environmental Sensing Technology and the Making of a Computational Planet* (Minneapolis, MN: University of Minnesota Press, 2016), pp. 57–80, https://doi.org/10.5749/minnesota/9780816693122.003.0003

14   Susan Schuppli, 'Dirty Pictures', in *Living Earth: Field Notes from the Dark Ecology Project 2014–16*, ed. by Mirna Belina (Amsterdam: Sonic Acts Press, 2016), pp. 189–210.

15   Jim Robbins, 'Oceans are Getting Louder, Posing Potential Threats to Marine Life', *New York Times* 21 January 2019, https://www.nytimes.com/2019/01/22/science/oceans-whales-noise-offshore-drilling.html

16   Karen Weintraub, 'These Whales are Serenaders of the Sea', *New York Times*, 7 January 2019, https://www.nytimes.com/2019/01/07/science/whales-songs-acoustics.html

Maybe the whole ocean is their [cetaceans'] auditory apparatus and archive; by joining their water-based inner ear with the outer ear of the ocean, *perhaps they have a medium for being in time that resembles our recording media but contrasts with the apparent instantaneity of our oral communication.* What is perhaps natural for them — nonlinear data access — is a matter of cultural techniques for us, and is only made by recording media.[17]

Peters, a philosopher of communication technology, has argued that since the appearance of technological media of distant communication in the nineteenth century, we have forgotten the idea that media is *primarily* natural. In *Marvelous Clouds*, he argues that air, water, ground, fire, light, clouds are the elemental media of communication for beings. His point of departure in the philosophy of communication is — to simplify radically — Marshall McLuhan's idea of 'the medium is the message'. However, Peters demonstrates that the two — medium and content — are not entirely separable and shape each other. In the example given above, the properties of water and the sonic capacities of the cetaceans must be considered together as the communication medium.

The natural and built environments are the medium of communication across time and space.

Following Peters' conceptualization, this composite environment is not only the container or infrastructure of historical or intertemporal communication. Rather, it also provisions beings in time. At the same time as the media provision beings, beings (i.e., content makers) 'read' and interpret these traces and signs to subsist, thus generating emergences. In the oceans, water is the principal medium for beings, but water is itself shaped by geo-atmospheric forces and even by the techniques of water beings. Peters asserts that 'Dolphins and whales' techniques shape the environment to enhance their techniques. Tuna for example, take advantage of vortices to propel themselves through the water at speeds much greater than would be predicted from their body size and strength, benefiting from hydraulic phenomena their swimming creates'.[18] Humans use their technological capabilities to record and reflect upon processes which, in turn, re-shape the environmental media in which they are borne.

---

17 My emphasis. John Durham Peters, 'Of Cetaceans and Ships', in *The Marvelous Clouds: Toward a Philosophy of Elemental Media* (Chicago, IL: University of Chicago Press, 2015), pp. 53–114, at p. 96.

18 Ibid., p. 88.

Let us explore the historical process as a medium. The historical is typically thought to be activity engaged by humans; but human-initiated activity is enabled and exceeded by chains of actors/actants, each of which may be linked in other chains. Historical time is the chain of events and activities, emergences and circulations of materials, practices and ideas which leave traces and records in their built and natural environment. The natural and built environment provides both the medium of flow and reshapes the messages channeled across time and space. Some of these sequences die out, others have *con-sequences* and take on new lives, and others return or are renewed, whether they are recognized or not.

The 'ontological turn' in social sciences has tended to break down the subject-object binary by attributing agency or quasi-agency to animals, organisms, instruments — for identifying, measuring, evaluating, or attaining — and, not least, the natural elements, in any consideration of a human undertaking. To be sure, this hardly displaces the fundamental condition of human knowledge for understanding these roles. But we might note that animals and organisms can and do process chains of action; and the great variety of human knowledge forms alerts us to ways in which sequences of action and knowledge are exceeded.

Historical processes may be indicated and evidenced through tiny traces such as Roy's circulating lock of hair, itself an adaptation of circulating religious reliquaries. These traces can carry historical forms and ideas across continents and time. Regarding material history, it is now the gargantuan nuclear and hydro-power projects and millions of miles of fiber-optic cables on the ocean-bed that generate possibly still more *con-sequences* and counter-finalities that historiography and scientists may or may not be able to capture.

The point of this exercise is to suggest that the medium of intertemporal communication has a natural ground that humans share with other organisms, albeit with different proportions or relations of the nature-culture dialectic of the evolving, 'living earth'.[19] The communicator, the message/communication, and the medium are interdependently creative, much as the ocean waters are the medium

---

19   See Brian Swimme and Mary Evelyn Tucker, *Journey of the Universe* (New Haven, CT: Yale University Press, 2011).

and co-creators of its flows with oceanic beings. It is not then entirely surprising that historical processes are expressed in temporal patterns that resemble patterns and interactivities that we find in oceanic flows. At a fundamental level, the historical process engaging humans is also natural. Processes generate other processes. Events impact, churn, and disperse; parts die, and other parts transform, transmute, and return. As such, the flow of oceanic waters is not just a model or even a metaphor. The historical process is continuous with it.

It was not till the second half of the eighteenth century that the notion of history as the property of a subject — a nation or civilization — in linear time, tunneling through the past into the future, became the dominant mode of temporal knowledge, first in northwest Europe and subsequently across the world in the twentieth century. The factors that combined to produce the linear conception of time in Europe are also complex; conceptions of religion, science, and the quest for global resources generated the temporal framework for a capitalist mode of endless accumulation.

The ideal of the conquest of nature as the means to achieve human satisfactions is the driving force of modern history as conceived by nation states. The nation states' apparatuses of knowledge production also set up — to put it in the barest terms — the opposition between history and nature. This cosmological reversal from the pre-modern era perhaps saw its greatest triumph when humans — particularly early modern Europeans — began to successfully navigate the deepest oceans. For most humans, oceans had presented the limits of their capacities. Well into the Renaissance, the Pillars of Hercules that marked the passage between the Mediterranean and the Atlantic Ocean were said to bear the warning *Non plus ultra*, 'nothing further beyond', cautioning sailors and navigators to go no further.

Over the last two hundred years, humans have begun to colonize the ocean itself. This colonization has been industrial in nature, battering the ocean by massive commercial traffic and fishing, nuclear testing, constant bombardment for oil and gas explorations and militarized island buildings, among other invasions. Not least, it is being strangled as the dumpsite of the terrestrial planet. To my mind, the modern idea of the conquest of nature, and the institutionalized and technologized modes of exhausting it is radically unprecedented. We now come face to face with the hubris that human history can destroy, negate, and transcend the medium of its sustenance.

I have argued that historical time should to be understood in the terms of natural processes, more than has been done heretofore. As such, historical time is continuous with the nature of oceanic flows. For this purpose, I have rendered the historiographical process that is uniquely human to the background as something to be grasped in relation to historical-oceanic time. Reflexive historiography remains powerful; after all, it is a necessary condition of the capacity of humans to dominate nature. Can this reflexivity be turned to develop another relationship to nature?

Merely reducing fossil fuel emissions cannot address our problems. The massively growing scale of consumption and destruction of natural materials and organisms for increased production has depleted and destroyed the earth and the oceans. The treadmill of ever-increasing consumption for profits and GDP growth embedded in the deeply institutionalized cosmology of our times demands change. Is there still a way to reconcile the creative capacities of humans with the limits of nature? The re-direction of historiographical knowledge to accord better with the nature of historical time and the sovereign planet would mark an important step towards it.

The powers of symbolic representation that gave humans the capacity to race ahead of all species in the evolution of life may well have met its match in the revenge of the oceans. The greatest threats to the human world today appear from the oceans, from rising sea levels to geo-atmospheric transformations of recognizable climate patterns. The question that arises today is the extent to which the Anthropocene, an era where human activity has the greatest influence on climate and the environment, will ravage the ocean and the degree to which the ocean will ravage us.

Bibliography

Duara, Prasenjit, *Rescuing History from the Nation: Questioning Narratives of Modern China* (Chicago, IL: University of Chicago Press, 1995), https://doi.org/10.7208/chicago/9780226167237.001.0001

— *The Crisis of Global Modernity: Asian Traditions and a Sustainable Future* (New York, NY: Cambridge University Press, 2015), https://doi.org/10.1017/cbo9781139998222

— *Sovereignty and Authenticity: Manchukuo and the East Asian Modern* (Lanham, MD: Rowman and Littlefield, 2003).

El-Geziry, T. M., and I. G. Bryden, 'The Circulation Pattern in the Mediterranean Sea: Issues for Modeller Consideration', *Journal of Operational Oceanography*, 3.2 (2010), 39–46, https://doi.org/10.1080/1755876x.2010.11020116

Gabrys, Jennifer, 'From Moss Cam to Spillcam: Techno-Geographies of Experience', in *Program Earth: Environmental Sensing Technology and the Making of a Computational Planet* (Minneapolis, MN: University of Minnesota Press, 2016), pp. 57–80, https://doi.org/10.5749/minnesota/9780816693122.003.0003

Goody, Jack, *The Theft of History* (New York, NY: Cambridge University Press, 2006), https://doi.org/10.1017/cbo9780511819841

Lozier, M. Susan, 'Deconstructing the Conveyor Belt', *Science*, 328.5985 (2010), 1507–11, https://doi.org/10.1126/science.1189250, http://science.sciencemag.org/content/328/5985/1507

McNeill, John, 'Historians, Superhistory, and Climate Change', in *Methods in World History: A Critical Approach*, ed. by Janken Myrdal, Arne Jarrick, and Maria Wallenberg Bondesson (Lund: Nordic Academic Press, 2016), pp. 19–43, https://doi.org/10.21525/kriterium.2.b

Peters, John Durham, 'Of Cetaceans and Ships', in *The Marvelous Clouds: Toward a Philosophy of Elemental Media* (Chicago, IL: University of Chicago Press, 2015), pp. 53–114.

Raj, Kapil, 'Beyond Postcolonialism… and Postpositivism: Circulation and the Global History of Science', *Isis*, 104.2 (2013), 337–47, https://doi.org/10.1086/670951

Robbins, Jim, 'Oceans are Getting Louder, Posing Potential Threats to Marin Life', *New York Times*, 21 January 2019, https://www.nytimes.com/2019/01/22/science/oceans-whales-noise-offshore-drilling.html

Schuppli, Susan, 'Dirty Pictures', in *Living Earth: Field Notes from the Dark Ecology Project 2014–16*, ed. by Mirna Belina (Amsterdam: Sonic Acts Press, 2016), pp. 189–210.

Slif, Brent D., *Time and Psychological Explanation* (Albany, NY: SUNY Press, 1993).

Swimme, Brian, and Mary Evelyn Tucker, *Journey of the Universe* (New Haven, CT: Yale University Press, 2011).

Thoreau, Henry David, 'Resistance to Civil Government', in *Aesthetic Papers*, ed. by Elizabeth Peabody (New York, NY: G. P. Putnam, 1849), pp. 189–213.

Tucker, Mary Evelyn, and John Grim, 'Four Commentaries on the Pope's Message on Climate Change and Income Inequality', *The Quarterly Review of Biology*, 91.3 (2016), 261–70.

Weintraub, Karen, 'These Whales are Serenaders of the Sea', *New York Times*, 7 January 2019, https://www.nytimes.com/2019/01/07/science/whales-songs-acoustics.html

Whitehead, Alfred North, *Process and Reality*, ed. by David Ray Griffin and Donald W. Sherburne (New York, NY: Free Press, 1985).

8. Affectual Insight

Love as a Way of Being and Knowing

David L. Haberman

There is a poetic phrase commonly known in the north Indian region of Braj: *pritama prita hi te peyi*, which means 'the Beloved is found precisely through love'. I would like to meditate on this seemingly simple sentence for the deeper meaning it has to offer our considerations of multiple ways of being and knowing, particularly as they relate to the living Earth community.

The Krishnaite traditions of Braj are informed by early Hindu Vedantic texts that give philosophical expression to an understanding of all life as simultaneously radically unified and bountifully diversified. This is recognized as the *siddantik* perspective that provides the philosophical basis for understanding the true nature of all life. Since accounts of creation are productive points of entry into the general worldview of a particular tradition, I begin with the emergent understanding of creation that is encountered in the foundational Vedantic texts, the Upanishads. One of the major tenets of Vedantic religious philosophy is non-duality (*advaita*). Principal Upanishadic texts recount that in the beginning the One Ultimate Reality (*Brahman, atman, purusha*) was lonely and bored, as there is not much joy in playing with one's self alone (*Brihadaranyaka Upanishad* 1.4).[1]

Hence, the One desired others, and, as a result, divided itself and proceeded to interact with itself in a multitude of manifest forms (*nama-rupa*). Through this creative process, the unmanifest One

---

1   For an English translation of this text, see: *Upanishads*, trans. by Patrick Olivelle (New York, NY: Oxford University Press, 1996), pp. 13–14.

    https://doi.org/10.11647/OBP.0186.08

produced out of itself the manifold world of all entities, both animate and inanimate. Accordingly, everything in the world is concurrently non-different and different, one and many. Although difference is acknowledged — and even celebrated — sharp ontological divides do not exist. It is all *Brahman*, for the fundamental Upanishads declare: 'this whole world is *Brahman*'. Significantly, the sacred totality includes a manifest (*adhibhautik*) as well as an unmanifest (*adhyatmik*) dimension. In contrast to, for example, a Calvinistic view, there are no impenetrable boundaries between the supreme reality (*Brahman* or God) and the visible world. As it says in the important Vedantic text the *Bhagavad Gita*, 'God is everything' (7:19).[2] Therefore, everything in the world is a part of the ultimate Whole, and, as such, is sacred. Animated presence pervades everything. The religious studies scholar Emma Tomalin calls this 'bio-divinity', the notion that Nature is infused with animated divinity, an idea widespread in India for a very long time.[3] An intellectual understanding of the *siddhantik* philosophical assertion that a sacred presence pervades everything, however, is not enough; it must be *realized*. And this leads us to another important perspective: the *bhavatmik*.

Bhavatmik is an adjective meaning the perspective 'whose essence is *bhava*'. The key term in this compound is *bhava*, a complex Sanskrit word with multiple dictionary meanings that include a state of mind, manner of being, way of thinking or feeling, emotion, attitude, affection, disposition, or realization.[4] In the Braj Krishnaite traditions, this word is best and most simply translated as 'love'. It is this term I had in mind in assigning the title to this short essay. A common way of expressing the goal of the Braj religious traditions is with the word *sarvatma-bhava*, which I would render as 'a loving realization of the divine in everything'. The crucial question is: how does one actually come to know the sacred or divine presence in some living entity? I return to the poetic phrase that I began with; the answer is precisely

---

2   My translation of the Sanskrit '*Vasudevah sarvam*' from the Bhagavad Gita 7.19. For the Sanskrit original with a readable English translation of this text see Winthrop Sargeant, *Shri Bhagavad Gita* (Albany, NY: State University of New York Press, 1993).

3   Emma Tomalin, 'Bio-Divinity and Biodiversity: Perspectives on Religion and Environmental Conservation in India', *Numen*, 51.3 (2004), 265–95, https://doi.org/10.1163/1568527041945481

4   See R. S. McGregor, *The Oxford Hindi-English Dictionary* (Delhi: Oxford University Press, 1993), pp. 765–66.

through love. When someone approaches us with aggression, shouting with rage in our face, we tend to pull back protectively and conceal ourselves. On the other hand, when someone approaches us with tender love, we are drawn out and reveal much more of ourselves. Might it be this way with all entities?

Conversations with religious practitioners in Braj suggest that it might indeed. There is a common thread that runs through hundreds of interviews I have had with worshippers of sacred rivers, trees, and mountains in northern India. Many reported that what formerly seemed like an 'ordinary' river, tree, or mountain revealed its divine nature or true sacred form (*svarupa*) after a period of interacting with that entity through loving acts of worship (*seva*). I met a young man on the bank of the Yamuna after watching him perform a very moving worship of the river. He explained to me how he had come to this: 'I used to see Yamuna-ji as an ordinary river and treat it badly. But then I met my guru, and he told me to start worshiping Yamuna-ji. At first I was a little resistant, but I did what he said. Soon, I began to see her *svarupa* (true divine form) and realized how wonderful (*adbhut*) she really is. So now I worship her everyday with love. The main benefit of worshiping Yamuna-ji is an ever-expanding love. I want to live in her world of love'.[5] This man suggests something very important. Loving attitudes and actions lead to a perspectival awakening; through a reverent approach one comes to know the true nature of some entity in a manner that exceeds mere intellectual knowledge. Once that true nature is revealed and one has an experience of its marvelousness, one enters spontaneously into an appreciative and worshipful attitude, and engages naturally in acts of loving care.

This is a common story. A woman I met who lovingly revered a neem tree daily with worshipful acts of service told me the result of this was that 'Mother revealed herself to me, which has led me to a very close relationship with her. I cannot now ever imagine cutting a living tree'.[6] A woman who worships a stone from Mount Govardhan in her home

5   David L. Haberman, *River of Love in an Age of Pollution: The Yamuna River of Northern India* (Berkeley, CA: University of California Press, 2006), p. 185.

6   David L. Haberman, *People Trees: Worship of Trees in Northern India* (New York, NY: Oxford University Press, 2013), https://doi.org/10.1093/acprof: oso/9780199929177.001.0001. Selected from additional notes from ethnographic interviews with tree worshipers for this book project.

every day reported that after some time, this stone revealed its *svarupa*, or true nature, and that this has led to a deep relationship in which the stone sometimes even talks to her. 'Before I began my worship, I never imagined how truly wonderful these stones were'.[7] During my visits to an outdoor shrine housing a stone from Mount Govardhan, a man I spoke with explained: 'When people lovingly decorate a Giriraj *shila* (Govardhan stone) and worship it, the personality comes out. Look, there are many Giriraj *shila*s here, but the *svarupa* is really showing itself in this one (*svarupa nikal deta hai*) because people have added eyes and decorations, and have worshiped it'.[8] Love is both a way of acting and an emotional state of being, and loving attitudes and actions are the very doorway into an insightful world of realization; they are concrete levers for opening up new perspectives. This is what I mean by 'affectual insight'. Many claim that through love the face of the Beloved is available in every entity.

This sense of the value of loving relationality has also been underscored by some important western scientific thinkers. For example, the 1983 Nobel Prize winning biologist Barbara McClintock promoted a 'feeling for the organism' as a crucial element in knowing it.[9] She called herself a 'mystic in science' and endorsed a form of attention based on loving relationship as a way of seeing things not available to the more aggressive approaches represented by such figures as Francis Bacon. In seeming agreement, Norman Brown promotes Alfred North Whitehead's notion of 'a science based on an erotic sense of reality rather than an aggressive, dominating attitude towards reality'.[10]

How does one develop a loving connection with the living Earth community that leads to affectual insight into its true nature or divine presence (*svarupa*)? Though the whole world is sacred, human beings are not good at connecting with abstract universalities. We are embodied beings designed to connect with tangible particularities. Our

7   David L. Haberman, *Loving Stones: Making the Impossible Possible in the Worship of Mount Govardhan* (New York, NY: Oxford University Press, 2020), p. 210.
8   Ibid., p. 208.
9   See Evelyn Fox Keller, *A Feeling for the Organism: The Life and Work of Barbara McClintock* (New York, NY: Henry Holt and Co., 1984). The phrase is found throughout the verbal expressions of McClintock. See, for example, p. xxii.
10   Norman O. Brown, *Life Against Death: The Psychoanalytical Meaning of History* (Middletown, CT: Wesleyan University Press, 2013), p. 316.

nature is such that our most powerful relationships are with specific, individual beings. Universal love, for example, is a noble sentiment, but it cannot begin to compare to the passionate engagement with the intimate love of a particular person. Intimate interaction with natural entities in northern India tends to be directed toward individual trees, rivers, and stones. There is an element of personal possession (*mamata*) in matters of love, as 'this is *my* child' or 'here is *my* lover'. A woman who maintains a Govardhan stone shrine in her home confirmed this viewpoint with a familiar example: 'There are many men in the world who are husbands, but the one who lives in this house is *my* husband. Likewise, there are many Giriraj *shila*s, but this one (gesturing to the Govardhan stone in her home shrine) is *mine*'.[11] But as love of a particular matures with concomitant knowledge, the scope of that love tends to broaden, just as people with no prior regard for dogs tend to look at dogs differently once they become friends with one particular dog.

The possibility of reverent interaction with a particular entity opening up to a more universal reverence was highlighted for me during an instructive conversation. One day I visited a large peepul tree shrine in Varanasi, and there I met a woman who was a *sadhvi*, a female practitioner who had renounced ordinary domestic life to devote herself to spiritual pursuits. At one point in our conversation, she explained what she thought was the real value of worshiping a tree: 'From the heartfelt worship of a single tree, one can see the divinity in that tree and feel love (*bhava*) for it. After some time, with knowledge one can then see the divinity in all trees. Really, in all life. All life is sacred because God is everywhere and in everything. This tree is a *svarupa* of Vasudeva (Krishna). As it says in the *Bhagavad Gita*, from devotion to a *svarupa* (one's own particular sacred form of God) comes awareness of the *vishvarupa* (universal form of God)'.[12] In brief, this knowledgeable woman was advancing the idea that the love of a particular has the possibility of opening up a more reverent attitude toward the universal. Regarding trees, her point was that loving interaction with a particular tree could lead to the realization of the sacrality of all trees — and by extension, of all life. With the

---

11   Haberman, *Loving Stones*, p. 198.

12   Haberman, *People Trees*, p. 197.

comprehension of the universal via the particular, we return full circle to the notion of *sarvatma-bhava*, the idea that everything is sacred. What first began as a proposition, is now deeply realized through affectual experience.

I close by suggesting that a new and special kind of love is available to us during these challenging times; a possibility is being offered to us that has perhaps never existed before. This may be the great redeeming feature of this troubled age of massive extinction, the silver lining in the proverbial dark cloud of our times. It is a love that is both astonishingly sweet and extremely urgent. Earth is singing us a special love song, if we can only open our hearts to hear it.

Today, we have the possibility of loving the living Earth community, of loving old-growth forests for example, in a manner that has perhaps never been possible before, for we now experience them at once as overwhelmingly beautiful and as *tremendously vulnerable*. The powerful, vibrant forests that frightened the early Puritans when they landed on the eastern seaboard and led them to conquer and clear-cut them have now been replaced by forests vastly diminished — and perhaps even dying. Much of the success of the great work we are being called to depends on understanding the nature of and embodying this special love. Although as a lifelong student of religion I would insist that love is fundamentally unified, I want to assert that the love we seem to be called to today has a dual nature: it is a boundlessly joyful love, and it is an affectionately concerned love. It is somewhat similar to the love a parent feels for a child ill with a life-threatening disease: the parent feels deeply moved by the child's smile while simultaneously being aware that the disease may take the child before her time. The intensity of the love is increased by the vulnerability of the child, and now so much of the living Earth community is endangered. The joyful dimension of this love has to do with opening ourselves to a power and wondrous presence in the world beyond even our greatest knowledge. The caring dimension of this love has to do with being sensitively attentive to the needs of particular threatened forms of life. This insightful, tender love offers a pathway to a deeper way of knowing and a more sensible way of being in the world today. Long live the whole living Earth community!

Bibliography

Brown, Norman O., *Life Against Death: The Psychoanalytical Meaning of History* (Middletown, CT: Wesleyan University Press, 2013).

Haberman, David L., *Loving Stones: Making the Impossible Possible in the Worship of Mount Govardhan* (New York, NY: Oxford University Press, 2020).

— *People Trees: Worship of Trees in Northern India* (New York, NY: Oxford University Press, 2013), https://doi.org/10.1093/acprof:oso/9780199929177.001.0001

— *River of Love in an Age of Pollution: The Yamuna River of Northern India* (Berkeley, CA: University of California Press, 2006).

Keller, Evelyn Fox, *A Feeling for the Organism: The Life and Work of Barbara McClintock* (New York, NY: Henry Holt and Co., 1984).

McGregor, R. S., *The Oxford Hindi-English Dictionary* (Delhi: Oxford University Press, 1993).

Olivelle, Patrick, trans., *Upanishads* (New York, NY: Oxford University Press, 1996).

Tomalin, Emma, 'Bio-Divinity and Biodiversity: Perspectives on Religion and Environmental Conservation in India', *Numen*, 51.3 (2004), 265–95, https://doi.org/10.1163/1568527041945481

Sargeant, Winthrop, *Shri Bhagavad Gita* (Albany, NY: State University of New York Press, 1993).

9. Confucian Cosmology and Ecological Ethics

Qi, Li, and the Role of the Human

Mary Evelyn Tucker

In our search for more comprehensive and global ethics to meet the critical challenges of our contemporary situation, the world's religions are emerging as major reservoirs of depth and insight, particularly with regard to the pressing environmental crises of our times.[1] While the scale and scope of the crises are being debated, few people would deny the seriousness of what we are facing as a planetary community immersed in unsustainable practices of production, consumption, and development. Clearly the world's religions have some important correctives to offer in this respect.

There is a growing realization that attitudinal changes toward nature will be essential for creating sustainable societies, in addition to new scientific, technological, and economic approaches to our environmental problems. Humans will not preserve what they do not respect. What is currently lacking is a moral basis for changing our exploitative attitudes toward nature. We have laws against homicide, but not against ecocide or biocide. Thus, we are without a sufficiently broad environmental ethics to alter our consciousness about the Earth and our life on it.

---

1   This was one of the main objectives of the Harvard conference series and edited volumes on Religions of the World and Ecology (http://fore.yale.edu/religions-of-the-world-and-ecology-archive-of-conference-materials/). From that series, see Mary Evelyn Tucker and John Berthrong, eds, *Confucianism and Ecology* (Cambridge, MA: Harvard University Press, 1998); and John Grim and Mary Evelyn Tucker, *Ecology and Religion* (Washington, DC: Island Press, 2014).

   https://doi.org/10.11647/OBP.0186.09

Consequently, what should concern us is this: to what extent can the religious traditions of the world provide us with ethical resources and cosmological perspectives that can help us redefine mutually-enhancing human-Earth relations?

The dynamic and holistic perspective of the Confucian worldview may offer significant contributions in this regard, enlarging our sense of the ethical terrain and moral concerns, and providing a rich source for rethinking our own relationship with nature. Confucianism's organic holisms can give us a special appreciation for the interconnectedness of all life forms and renew our sense of the sacredness of this intricate web of life. Moreover, the Confucian understanding of the dynamic vitalism underlying cosmic processes offers us a basis for reverencing nature. From a Confucian perspective, nature cannot be thought of as simply composed of inert, dead matter. Rather, all life forms share the element of *qi* or material force. This shared psycho-physical entity becomes the basis for establishing a reciprocity between the human and nonhuman worlds.

In this same vein, in terms of self-cultivation and the nurturing of virtue, the Confucian tradition provides a broad framework for harmonizing human life with the natural world. An example of this is the Confucian view of the human as a child born out of the Universe and the Earth, and thus owing filial respect and reciprocity to the Earth community. Another example is the Confucian understanding of virtues as having both a cosmological and a personal component, so that love is a generative force expressed by humans, but is also seen as comparable to the original generative forces in the universe. Thus, nature and virtue, cosmology and ethics, knowledge and action are intimately linked for the Confucians in China, Korea, and Japan. This chapter will concentrate on three major themes and their implications for ecological ethics: *qi, li,* and the role of the human.

We are aware that, like all spiritual traditions, there is always a gap between aspiration and realization of these practices. Despite the narrow, ideological views of Confucianism, China is not a model of an ecologically realized society, historically or at present. However, aspects of this worldview are worth retrieving in order that we may break out of the constraining perspective of a modern reductionistic worldview.

Qi

The Chinese have a term to describe the vibrancy and aliveness of the universe. This is *qi* or *ch'i*, which is translated in a variety of ways in the classical Confucian tradition as spirit, air, or breath, and later in the Neo-Confucian tradition as material force, matter energy, vital force. It describes the realization that the universe is alive with vitality and resonates with life. What is especially remarkable about this ancient and enduring realization of the Chinese people is that *qi* is a unified field embracing both matter and energy. It is thus a matrix containing both material and spiritual life from the smallest particle to the largest visible reality. *Qi* moves through the universe from the constituent particles of matter to mountains and rocks, plants and trees, animals, and birds, fish and insects. All the elements — air, earth, fire, and water — are composed of *qi*. We humans, too, are alive with *qi*. It makes up our body and spirit as one integrated whole, and it activates our mind-and-heart, which is a single unified reality in Chinese thought.

In other words, *qi* courses though nature, fills the elements of reality, and dynamizes our human body-mind. It is the single unifying force of all that is. It does not posit a dichotomy between nature and spirit, body and mind, matter and energy. *Qi* is one united, dynamic whole — the vital reality of the entire universe.

The implications of this unified view of reality become apparent to us rather quickly. One wants to know and experience this *qi* more fully. This is why most of the martial arts and exercises like *taiqi* aim to cultivate and deepen *qi*. Humans, for all their obliviousness, are intelligent enough to want to taste and savor this marvelous aliveness of the universe. They want to harmonize their most basic physical processes with *qi* — thus the dynamic coordination of breath and movement is at the heart of the Chinese physical arts. And arts they are indeed — this is not just a physical toning of the body or building up of muscles. This is a spiritual exercise filled with potency for health of mind and body — a coordinated and aesthetically pleasing dance of the human system in and through the sea of *qi*.

One way to visualize *qi* is as a vast ocean of energy, an infinite source of vibrant potency, a resonating field of dynamic power that is *in* matter itself, not separate from it. For *qi* is matter-energy, material force. This is the important contribution of Chinese thought to world philosophy. It is an insight and realization of particular significance for our contemporary world, which has been broken apart by our Enlightenment separation of matter and spirit, of body and soul, of nature and life.[2]

From the perspective of *qi* the world is alive with a depth of mystery, complexity, and vibrancy that we can only begin to taste and never fully exhaust. The sensual world *is* the spiritual world from the perspective of *qi*. The dynamism of each particular reality begins to present itself to us — the oak tree in the forest radiates an untold energy, the snow-covered mountains in the distance are redolent with silent *qi*, the rivers coursing to the ocean are filled with the buoyancy of *qi*.

One of the earliest Confucian writers, Mencius, speaks of the great flood-like *qi*. This is what I am evoking here. We are flooded, surrounded, inundated by *qi*. We walk around completely unconscious most of the time that this ocean of energy is here — sustaining us, nourishing us, and enlivening us. *Qi* is the gift of the universe — the endlessly fecund life source unfolding before us and around us in a daily miracle of hidden joy. It is the restorative laughter of the universe inviting us into its endless mystery.

As we return to the Chinese sources to sift through the texts and commentaries, what becomes apparent is that the notion of *qi* is not constant, but evolving. Nor is it unified and consistent. It is rather a multivalent idea that begins to reveal something of its shape and function only when seen from a variety of perspectives and texts.

In the classical Confucian tradition, *qi* tends to refer more generally to the spirit which animates the universe, the breath which enlivens humans, and the air that connects all things. Even from its earliest articulation, however, *qi* was never seen as an entity apart from matter. Rather, it is embedded in the natural and the human world. It animates and nourishes nature and humans. Indeed, the very Chinese character itself is said to represent the steam rising from rice, suggesting the nourishing and transforming power of *qi*. Like food, *qi* maintains life and human energy. Benjamin Schwartz observes, 'The image of food even suggests the interchange of energy and substance between humans and

---

2   On the Enlightenment legacy, see also 'Introduction: Ways of Knowing, Ways of Valuing Nature' in this volume.

their surrounding environment'.[3] The idea of *qi* as having the properties of condensation and rarefaction like steam suggests the same.

As the later Han and Neo-Confucians began to articulate their cosmological understandings, the unity of *qi* as matter-energy became more evident. Dong Zhongshu (170–104 BCE), the leading Han Confucian, described *qi* as a 'limpid colorless substance' which fills the universe, 'surrounds man as water surrounds a fish', and unites all creation.[4] The Neo-Confucians, however, developed the notion of *qi* to refer to the substance and essence of all life. It pervades and animates the universe as both matter and energy.

For the Neo-Confucian, Zhang Zai (1020–77 CE), the vibrancy of material force originates in the Great Vacuity (*taixu*) which contains the primal, undifferentiated material force. As it integrates and disintegrates it participates in the Great Harmony (*taihe*) of activity and tranquillity. This perspective affirms the unified and real processes of change, not seeing them as illusory, as the Buddhists might, nor as a product of a dichotomy between non-being and being, as the Daoists would. There is, instead, a dynamic unity of *qi* as seen in its operations as both substance (emerging in the Great Vacuity), and function (operating in the Great Harmony).

Li

Li is the inner ordering principle of reality that is embedded in the heart of *qi*. The Chinese character for *li* suggests working on the geological veins found in the mineral jade, which must first be discovered, and then carved adeptly. *Li* is comparable to the principle of *logos*, whereby all of reality is imprinted with structure and intelligibility. It is both pattern and potential pattern, and thus gives reality its intricacy of design as well as its thrust toward directionality and purpose. It is a revealing and concealing sensibility for human consciousness. We seek to find its imprint in the flow of the natural world around us, as well as in the unfolding of our lives. As Thomas Berry often said, we have lost the ability to perceive this

---

3   Benjamin Schwartz, *The World of Thought in Ancient China* (Cambridge, MA: Harvard University Press, 1985), p. 180.
4   Wm. Theodore de Bary, ed., *Sources of Chinese Tradition* (New York, NY: Columbia University Press, 1960), p. 466.

vast intelligibility of the universe and thus have become ungrounded and rudderless, locked in our own self-referential mindsets.

It is, however, the universe which is calling to be read and to be heard in the deep patterning of its particularities. The beauty of *li* is that it brings us into contact with the myriad forms of life, the 'ten thousand things' (*wanwu*) as the Chinese say, with a penetrating clarity. This is because *li* is both normative principle and intelligible pattern. As pattern, it gives us entry into understanding nature and its complex workings. As principle, it gives us a grounding for a morality that arises from the very structure of life itself. The moral dimensions of the universe are found in the depths of matter revealing itself to us as *li*.

Li is principle and pattern — both a moral and a natural entity bringing together our profound embeddedness in a universe of meaning and mystery. The allure of the universe lies in seeing and experiencing that meaning and mystery before us, behind us, and all around us. We are drawn forth into a sense of the breadth and depth of *li* as manifest in the phenomenal world in great diversity and particularity. All of this breadth and depth of inner ordering is gathered up in the Great Ultimate (*taiji*) — that which contains and shapes and generates all principles and patterns in the universe.

As one of the principal Neo-Confucian thinkers, Zhu Xi (1130–1200) says, 'The Supreme Ultimate is merely the principle of Heaven-and-Earth, and the myriad things'.[5] According to another leading Neo-Confucian, Cheng Yi (1033–1107), 'Principle is one (in the Great Ultimate); its manifestations are many (in the world)'.[6] To illustrate this, both use an analogy involving the moon shining in the water in the irrigated rice fields on a terraced mountain side. There are many moons which are reflected, but only one full moon in the sky. *Taiji* is like this full moon. It is translated as the Great Ultimate or the Supreme Ultimate, while the term itself refers to a pole star — guiding, illuminating, and alluring. For Cheng Yi, and his brother Cheng Hao (1032–85), *li* was like a genetic coding, and was thus identified with the creative life principle (*shengsheng*).[7]

---

5   Ibid., pp. 701–02.
6   Wm. Theodore de Bary and Irene Bloom, eds, *Sources of Chinese Tradition* (New York, NY: Columbia University Press, 1999), p. 700.
7   Ibid., p. 689.

The creative dynamics of this great container of principles are cosmological, namely there is an interaction of non-being and being or the unmanifest and the manifest. This is seen in the interaction of the *wuji* (Non-ultimate) and the *taiji* (Great Ultimate). Some of the most interesting arguments and discussions in Chinese thought have arisen among thinkers who are commenting on this complex interaction.

Some would say that the Daoists want to maintain a dichotomy between non-being and being, emphasizing the dynamic creativity of non-being as the source of all life. Others would say the Buddhists want to maintain the ultimate emptiness of non-being and the illusory quality of being. The Neo-Confucians struggled to assert the importance of the dynamic continuity between these two forces (non-being and being). Indeed, they would maintain that the very creativity of the universe is revealed in this dialectical interaction. The complementarity of these creative forces is at the heart of all cosmological processes for the Confucians. The vast changes and transformations of nature in the endless flow of *qi* become clear in this interaction. That is because all reality, namely all *qi*, is imprinted with *li*. Discovering this patterning in the fluid material force of the universe is the challenge for humans.

As *li* is unveiled, humans can discern the appropriate patterning for both their individual and their collective lives. The universe unfolds according to these patterns of deep structure embedded in reality. Social systems are established according to these patterns, agriculture is conducted in harmony with these patterns, politics functions in relation to these patterns, and individuals cultivate themselves in response to these patterns.

The Role of the Human

In the Neo-Confucian understanding, humans receive *li* from heaven. Their heavenly endowed nature is thus linked to the patterning throughout the universe. By the same token, humans are composed of *qi*, the same dynamic substance that makes up the universe.

Humans are thus imprinted with unique and differentiated *li* embedded in *qi*, the material force of their own body-mind. *Li* guarantees the special and different qualities of each human being, while *qi* establishes the material and spiritual grounds for subjectivity, thus uniting humans with one another and with the vast world of nature. In other words, *qi* as vital force is the interiority of matter, providing the matrix for communion and exchange of energy between all life forms.

Humans, then, are given a heavenly endowed nature which joins them to the great triad of Cosmos, Earth, and other humans. While this heavenly endowed nature is a gift of the universe from birth, it is understood as something to be realized over a lifetime. This realization of one's full nature occurs through the process of self-cultivation, which is at the heart of Confucian moral and spiritual practice. This process of actualization is not abstract or otherworldly, but rather concerned with the process of becoming more fully human. In doing so, one penetrates principle and perceives pattern amidst the flux of material force in ourselves and in the universe at large. The goal of our cultivation is to actualize and recognize the profound identity of ourselves with heaven, Earth, and the myriad forms of life.

Because the *qi* that we are each given may vary in its purity or turbidity, cultivation is needed. Evil, imperfection, loss, and suffering are thus part of the human condition. The Confucians, however, believe one's heavenly endowed human nature is essentially good and thus perfectible. To illustrate this, Mencius uses the example of a child about to fall into a well (*Mencius* II A 6). The instinct of any person is to save the child from harm, not for any exterior reasons but due to a naturally compassionate heart. The key to the goodness of human nature is a profound sympathy or empathy which all humans have. Indeed, affectivity is what distinguishes humans in the Confucian worldview. As Mencius says, 'No one is devoid of a heart sensitive to the suffering of others' (*Mencius* II A 6). Because of this basic sympathy, Confucians affirm that at the level of our primary instincts we will tend toward the good. Mencius uses wonderfully evocative images from nature to illustrate this, like water flowing naturally downhill (*Mencius* VI A 2). Like wind blowing over grasses, people are inclined toward the good and respond to the good because they are imprinted with the good.

From these examples, Mencius goes on to describe the basic seeds implanted in human nature which, when cultivated, become the key virtues for living a fully humane life. The seeds are compassion, shame, courtesy, and modesty, and a sense of right and wrong (*Mencius* II A 6, *Mencius* VI A 6). These seeds need to be watered and tended so that they grow and flourish into the primary Confucian virtues of humaneness, righteousness, propriety, and wisdom. The images used to describe the growth and cultivation of virtue are derived from the agricultural patterns and seasonal cycles of humans dependent on nature. Consequently, I am inclined to use the metaphor of 'botanical cultivation' when speaking of Confucian moral and spiritual practice.

The aim of such practice is to allow the seeds or tendencies of our deepest human spontaneities to be nourished and to flourish. Mencius suggests that this should be as clear as tending trees in one's garden: 'Even with a *tong* or a *zi* tree one or two spans thick, anyone wishing to keep it alive will know how it should be tended, yet when it comes to one's own person, one does not know how to tend it. Surely one does not love one's person any less than the *tong* or the *zi*' (*Mencius* VI A 13). In this same spirit, there should develop a naturalness to our actions based on the rhythms of the cosmos itself. From seeds in the soil to seasons and their cycles, to the flow of rivers and the thrust of mountains, we are part of the rhythms of the universe and need to nourish our original nature.

If one develops these seeds, it is like 'a fire starting up or a spring coming through'. The moral power that results from this cultivation of virtue is boundless: 'When these (seeds) are fully developed, one can take under one's protection the whole realm within the Four Seas, but if one fails to develop them, they will not be able even to serve one's parents' (*Mencius* II A 6).

The key is to tend, to activate, and to align our deepest spontaneities with the dynamic patterns of change and continuity in nature. Thus, self-cultivation needs to be an organic process. As Mencius suggests, we need to nourish our flood-like *qi* with integrity (*Mencius* II A 2) and recover our original mind-and-heart (*Mencius* VI A 11). However, this cannot be a forced or artificial process. Mencius uses the example of the man from Sung who planted rice seedlings. In his desire to see them grow quickly, he pulled at them too soon, and they withered. As Mencius observes, 'There are few people in the world who can resist the urge to help their rice plants grow' (*Mencius* II A 2). Others leave them unattended or do not bother to weed. How to nurture and nourish is the art of cultivation in both nature and in humans.

Mencius also uses the example of Ox Mountain, where, due to deforestation and overgrazing, the mountain becomes denuded (*Mencius* VI A 8). Erosion sets in, and the ecosystem is destroyed. People are inclined to think this has always been the nature of the mountain. Improper cultivation of ourselves and of the land results in waste and loss. As Mencius says, if one is not restored by the natural rhythms of the day and night, but rather dissipates one's energies and becomes dissolute, people will think that dissolution is one's essential nature. However, he insists that nourishment is the key: 'Given the right nourishment there is nothing that will not grow, and deprived of it there is nothing that will not wither away' (*Mencius* VI A 8).

These examples are so simple, clear, and timeless. They are as appropriate for our day as for Mencius', as their natural imagery restores us to the deeper rhythms of our being in the universe. For, in this context, self-cultivation does not lead toward transcendent bliss or otherworldly salvation or even personal enlightenment. Rather, the goal is to move toward participation in the social, political, and cosmological order of things. The continuity of self, society, and cosmos is paramount in the Confucian worldview.

Thus, self-cultivation is always aimed at preparing the individual to contribute more fully to the needs of the contemporary world. For the Confucians, this implies a primacy of continual study and learning. From this perspective, education is at the heart of self-cultivation. This is not simply book learning or scholarship for the sake of careerism. It is rather education — leading oneself out of oneself into the world at large. More than anything, then, the role of the human is to discover one's place in the larger community of life. And this community is one of ever expanding and intricately connected concentric circles of family, school, society, politics, nature, and the universe. We are embedded in a web of relationships and one fulfills one's role by cultivating one's inner spontaneities so that one can be more responsive to each of these layers of commitments.

For the Confucians this is all set within the context of an organic, dynamic, holistic universe that is alive with *qi* and imprinted with *li*. Thus, finding one's role is realizing how one completes the great triad of heaven and Earth. As we rediscover our cosmological being in the macrocosm of things, our role in the microcosm of our daily lives will become more fulfilling, more joyful, more spontaneous. The pace and rhythm of our lives will be responsive to the rhythms of the day, the changes of the seasons, and the movements of the stars. The great continuity of our being with the being of the universe will enliven and enrich our activities. By attuning ourselves to the patterns of change and continuity in the natural world, we find our niche.

We thus take our place in the enormous expanse of the universe. We complete the great triad of heaven and Earth and participate in the transforming and nourishing powers of all things. In so doing, we will cultivate the land appropriately, nurture life forms for sustainability, regulate social relations adeptly and fairly, honor political commitments for the common good, and thus participate in the great transformation of things. This will be manifest as our own inner authenticity resonates with the authenticity of the universe itself.

This holistic and dynamic understanding of the world and the role of humans, found in Confucianism, could bring us far in the revisioning that is needed for us to deal effectively with our current ecological crisis. In turn, it is but one example of the potential benefit of tapping the resources of the world religions in our endeavor to formulate a more comprehensive and global ethics.

Bibliography

de Bary, Wm. Theodore, *Sources of Chinese Tradition* (New York, NY: Columbia University Press, 1960).

de Bary, Wm. Theodore, and Irene Bloom, eds, *Sources of Chinese Tradition* (New York, NY: Columbia University Press, 1999).

Grim, John, and Mary Evelyn Tucker, *Ecology and Religion* (Washington, DC: Island Press, 2014).

Lau, D. C., trans., *Mencius* (New York, NY: Penguin Books, 1970).

Schwartz, Benjamin, *The World of Thought in Ancient China* (Cambridge, MA: Harvard University Press, 1985).

Tucker, Mary Evelyn, and John Berthrong, eds, *Confucianism and Ecology: The Interrelation of Heaven, Earth, and Humans* (Cambridge, MA: 1998).

SECTION IV

STORYTELLING: BLENDING ECOLOGY AND HUMANITIES

Fig. 6 Central Californian Coastline, Big Sur. Photo by David Iliff (2017), Wikimedia, CC BY-SA 3.0, https://commons.wikimedia.org/wiki/File:Central_Californian_Coastline,_Big_Sur_-_May_2013.jpg#/media/File:Central_Californian_Coastline,_Big_Sur_-_May_2013.jpg

10. Contemplative Studies of the 'Natural' World

David Haskell

What might be learned by paying repeated, open-ended attention to particular places? Such acts of ecological meditation might reveal truths that complement and feed those uncovered by scientific experimentation, theoretical analysis, imaginative exploration, and creative expression.

I undertook my own exploration of this question in part as a response to the dislocations of academic culture. I was trained to use 'study systems' to answer questions of general interest. Questions from theory drove my engagement with living communities. Novelty was valued over familiarity: funding agencies and journals rewarded new ideas, not the process of listening to what was, on the surface, already known. Seeking new insight and knowledge is part of the restless exploratory spirit of science. These explorations reveal much that is useful, beautiful, and life-affirming. Yet, question-driven, agency-sanctioned research on a short-term funding cycle can reveal only part of what there is to know in this world.

And so: an experiment in observation. I set aside the next grant application and walked into the forest. I selected a rock flat enough to sit on for a while, and then started my watch. A circle of forest, about one square meter in area. I brought a hand lens and a notebook. I left behind the technologies — cameras, DNA sequencers, sound recorders, laptops — that were so helpful to me in other parts of my work. I also left behind hypotheses and lesson plans. Such focusing devices are useful, but they pre-determine and narrow our attention when we walk into a forest. Instead of a question to guide what I discovered, I brought

    https://doi.org/10.11647/OBP.0186.10

a commitment to return to a particular place and try to pay attention. I returned, over hundreds of visits, through a year. Then, over many years. Later, I used a similar practice with individual trees located in disparate places around the world. Show up; listen; repeat. This is, of course, an application of meditative practice in an ecological context.

My books relate the stories that I discovered in each place. Here I share a few general thoughts about the nature of these experiments in observation. I make no claim for novelty in either the practice or what I learned. Indeed, the fetish of intellectual or artistic innovation, so central to how we value ourselves and others in academia and the world of culture, is perhaps best set aside as we enter a contemplative practice. If novelty does eventually emerge, it will perhaps be the surprise of apprehending the ordinary in ways that are new to us or, later, the opening of new opportunities to share such familiarity with others.

Opening the Senses: Finding and Telling Stories of Place

We, like all animals, are covered in neuronal sensors. Most of these report data that are irrelevant to the task at hand at any given moment. So, the brain ignores or actively suppress most sensory input, focusing on the few stimuli that our neural circuits deem most relevant. Even these 'relevant' stimuli are filtered by our brain's limited capacity to receive and analyze information. The Ford truck engines, bicycle tires on concrete, ambulance sirens, house finch warblings, and wind smattering in cottonwood leaves outside my window were not present to me as I wrote this paragraph. This acoustic richness was blocked from my conscious awareness until I reached out to find them. Nowadays we are also surrounded by stimuli designed by humans and our computer algorithms to further capture, redirect, and narrow the scope of our attention. 'Opening the senses' is therefore a process that must countervail the distractions and filters imposed by both human ingenuity and animal evolution. Physical and psychological well-being demands a vigorous commitment to ignoring stimuli. In a world where closing off is often necessary, we perhaps also need spaces in which our deliberate practice is to re-open. Choosing where this opening will happen is the discipline of directed attention.

Start, perhaps, with a sensory inventory in a familiar space outdoors. Pick a sense, hearing, for example. Pour your attention into your ears. Name the sound. Return your attention to your ears. Name the sound that was behind the first. Then the next. Each time gently return your attention to the ears. Do the same, but don't name, just describe each sound, find words for the physicality of each aerial vibration. Then again, listen, repeatedly, but no names and no descriptions, just presence in the sound.

Within the first twenty seconds of the inventory, the richness of the sensory world becomes apparent. Thousands of sense impressions reveal themselves every minute. The richness of our inner world also emerges, sometimes forcefully. Our mind wanders to some other place, a place of distraction, or emotional energy, or boredom. We acknowledge the mind's wandering nature, then return to the inventory of the exterior world.

From this sensory treasury we gain two gifts.

First, our curiosity and creativity are fed. Stories of place, previously hidden by inattention, emerge. Here is where intellect and creativity can intersect with the world. Contemplative practice is not the opponent of intellectual inquiry and artistic expression, rather it is one method by which interior processes can transcend their limitations, then integrate with understandings that originate not in the self, but in the community of life.

Second, we can tell the stories of our place. The process of writing or of oral storytelling transports one human mind into the experience of another. Such a feat requires some help, especially if the transportation is to be convincing and emotionally powerful. Attending to the sensory particularity of place, then faithfully and vividly reporting this particularity is a helpful skill for any storyteller. Contemplative practice, seemingly so inward, is in fact a preparation for interconnection with other people.

Integration: Attending to Interconnection

In my work as a scientist and a teacher, I bring to every place a set of hypotheses or pedagogical goals. These focusing devices are powerful and necessary. But they also emerge from particular places in the structures of knowledge that we use to understand and then communicate the nature of the world. Walking into the forest to gather data on the evolutionary genetics of endemic snails is a different walk than one that measures the relative abundance of understory plants. Both require me to arrive at the forest with pre-determined narratives. The same is true of the various field labs that I lead with students. In each of these cases, we enter the forest not to discern which stories are most powerful on that day in that place, but to bring a story, embodied in our questions, to the forest. What we bring is, at best, a simplification of the interconnected processes of the world and, sometimes, merely a projection of unrooted theory.

The two practices — focused questioning and open-ended attention — are not in opposition to each another. Rather, reciprocal nourishment happens. Our institutional and intellectual structures, though, strongly favor targeted questioning and offer little encouragement for repeated attention to place. Within higher education, especially in the sciences, there are few if any mechanisms to help us deepen this latter practice, few formal structures or communities of colleagues from whom to learn and be guided.

Greater emphasis on using and refining the practice of open-ended attention might yield several benefits. Foremost among these is a counterweight to disciplinary specialization. Tapering and deepening of professional expertise are important and fruitful processes. A complementary practice of opening to the integrative insights of contemplative practice yields different kind of focus: a tapering and deepening of relationship to place rather than to a set of ideas and questions.

Limits

Extended contemplative engagement with place gave me a thorough schooling in my limitations. This was an education in the structure of knowledge, a spur to further investigation, and an invitation to humility.

Limits became apparent first through my senses. Returning again and again to each sense, I was drawn into the subtleties and fine details, encountering organisms and their signs on the edge of perception: tiny insects and nematodes undetectable without peering closely at the leaf litter. Mats of fungal strands that disappear into the soil as they ramify and flee the human eye. Odors of microbial life. Sounds of animals pushing against below-ground pebbles. The changing taste of the air through a morning, a reminder of the chemical-microbiological milieu in which we swim. These limitations of my unaided senses were a bodily reminder of human evolutionary ecology, each sense tuned by natural selection to the human niche. These sensory limits then remind me of more profound chasms. Even the most sophisticated microscopes and sequencers give only a partial sketch of life's community, a reminder of ignorance.

A sharp awareness of the limits of my knowledge followed on the heels of the lesson of sensory limitations. I started my forest observations after several decades of studying, researching, and teaching biology. Yet I was stunned by how little I knew of the forest community. Names for most species eluded me. The stories of each species were even more hazy. This was especially true for the very small creatures, the tiny diptera, fungi, and nematodes. Before I started this project, I thought I had a clear intellectual understanding that no-one can apprehend even part of a forest community. Days and days of sitting in this ignorance drove that understanding into my emotions and body. What was left was wonder at the unendingly fascinating and diverse world that we inhabit, mixed with frustration at my incompetence and dread as I took on the imposter's role of writing and teaching about a forest that will forever elude my understanding.

After my periods of observation in the forest, I went to the library to ferret through the primary literature. The forest gave me a reading list, a set of questions. One of my goals as a writer is to share this knowledge from the literature, to unveil and bring to life for non-specialists the remarkable findings of modern ecological and evolutionary biology. Here, too, limits are very clear. Many of the forest's actors are unknown and unnamed, and their roles and history unknown. These limits are being rapidly pushed back by the inquisitive energies of scientists, at least along the fronts deemed worthy by funding agencies. Nevertheless, I come away from the forest and the library with a humbling sense that our knowledge is vastly overshadowed by the enormity of life's diversity and the complexities of ecological stories. Our actions as citizens and members of life's community take place amid much ignorance.

The limits of our observational powers and the youthfulness of scientific understanding are limits that can, in principle, be overcome with technology and time. But even if knowledge were to be vastly expanded, epistemological limits would remain. Does a forest have inherent value? What is the most responsible way to live in relation to life's community? Is a particular management proposal good for the forest? Scientific data are needed to inform our answers to these questions, but data alone are insufficient. Ethical claims must draw on modes of understanding that include science but also transcend science's remit and abilities. Contemplative engagement within life's community may offer a bridge between science and ethics, as outlined below.

A Search, through Beauty, for Objective Foundations of Ethics

Where should we root our ethics? This is a vexed question for biologists. Contemplative observation of living communities might ease some of this trouble.

In an era of climate disruption, biotic impoverishment, and human injustice, finding a ground for ethical discernment and subsequent right action is critically important. Science, especially biological science, offers data that reveals and quantifies our problems. Science is also necessary to evaluate the relative technical merits of solutions. But science cannot tell us why to care or act. Indeed, some biologically-inspired philosophy claims that such ethical questions should be answered with nihilism: ethics is, in this view, a mental mechanism evolved by our ancestors, with no objective substance to beyond its effects on our nervous system. Ethics is thus posited as a neurobiological phenomenon, not a guiding imperative that exists beyond the human. This position seems incompatible with biodiversity conservation and environmental activism. These fields of study and action are founded on ethical imperatives. We should, they assume, protect life's community not only for its utilitarian value to humans, but for its own sake. This 'own sake' is meaningless if the ethical nihilists are right.

Theistic traditions answer this challenge by locating their ethical foundations in beings or relationships that exist partly beyond the natural order. Ethical nihilists and theists may be right, but I seek a foundation for ethics that is neither theistic or nihilist. I propose that contemplative engagement within the community of life may provide part of such a foundation. In *The Songs of Trees*, I build this case with argument, example, and metaphor.[1] I claim that participation in living communities leads to a mature sense of ecological aesthetics. This sense of beauty then serves as an integrative and partly objective foundation for ethics. In the following paragraph, I summarize the structure of the argument.

Life is made from networked relationships. We participate in these relationships at all times through our microbiome, diet, sub-cellular biochemistry, ecology, social connections, and culture. Contemplative observations of a particular place bring that network into awareness. Over extended periods of time, we come to understand and participate in the network through multiple modes: intellect, emotions, senses, microbiology, memory, and conversation with others. The deeper and longer our engagement, the more our body, mind, and emotions awaken to the strands of living network in which we live.

One result of this process of contemplative engagement is that our aesthetic sense matures. We come to understand what is broken or beautiful in our place in a deep, integrative way. We, in Iris Murdoch's words, partly 'unself' into the network, gaining a deep sense of beauty that transcends the 'self'.[2]

Aesthetics can, of course, mislead our ecological judgement through superficial sensory impressions. But in a mature sense of aesthetics, our brains draw together all we have learned from participation in the network into a sense of beauty that is integrative and far-reaching in its knowledge. This may serve as a (partly) objective foundation for ethical discernment.

In what way objective? In partly transcending the 'self' and entering into the experience of life's network, our sense of beauty and our judgement of the good are no longer properties solely of the individual. They reveal the nature of the network. Every 'node' or organism in the network of course has a different perspective and different sensory

---

1   *The Songs of Trees: Stories from Nature's Great Connectors* (New York, NY: Viking, 2017).

2   Iris Murdoch, *The Sovereignty of Good* (London: Routledge & K. Paul, 1970; repr., London: Routledge Classics, 2001), pp. 82–91.

abilities, but disparate parts of the network will converge on similar judgments, especially through a practice of opening to the network.

This view has four interesting consequences.

First, ethics are no longer confined to humans. Other species, in their own ways, can unself into networks and draw ethical conclusions. The more extensive the unselfing, the deeper the ethical insight.

Second, ethics are participatory. Sound ethical judgement emerges from lived experience within living networks. The intellect alone is insufficient.

Third, networked relationship relies on attention and conversation. We will reach better conclusions through listening to life's community, other humans included, a process that is often absent from our divided and siloed discourse about questions of ethical importance.

Fourth, beauty is an important guide and should therefore be re-elevated in our esteem, given attention in our educational systems, and its value cultivated through professional institutions. Although experiences and discussions of beauty are rarely included in science curricula or policy processes, the human animal continues to be deeply moved by beauty and our decisions strongly shaped by aesthetic choices. Unguided and shallow aesthetic judgments are dangerous, but mature and experienced aesthetic judgements are sources of great insight.

Contemplative experience can integrate aesthetics, ethics, and science — an integration acting as an antidote to institutional, educational, and psychological fragmentation. This is not surprising: contemplation of a forest, a stream, a city block shows us the integrated nature of the life's community and thus of our own being.

Bibliography

Haskell, David, *The Forest Unseen: A Year's Watch in Nature* (New York, NY: Viking, 2012).

— *The Songs of Trees: Stories from Nature's Great Connectors* (New York, NY: Viking, 2017).

— 'Listening to the Thoughts of the Forest', *Undark: Truth, Beauty, Science*, 7 May 2017, https://undark.org/article/listening-to-the-thoughts-of-the-forest/

Marois, René, and Jason Ivanoff, 'Capacity Limits of Information Processing in the Brain', *Trends in Cognitive Sciences*, 9.6 (2005), 296–305, https://doi.org/10.1016/j.tics.2005.04.010

Murdoch, Iris, *The Sovereignty of Good* (London: Routledge, 1970; repr., London: Routledge Classics, 2001).

Ruse, Michael, 'Evolutionary Ethics. A Defense,' in *Biology, Ethics, and the Origins of Life*, ed. by Holmes Rolston (Boston: Jones and Bartlett, 1995), pp. 89–112.

Tremblay, Sébastien, et al., 'Attentional Filtering of Visual Information by Neuronal Ensembles in the Primate Lateral Prefrontal Cortex', *Neuron*, 85.1 (2015), 202–15, https://doi.org/10.1016/j.neuron.2014.11.021

11. Science, Storytelling, and Students

The National Geographic Society's On Campus Initiative

Timothy Brown

In the winter of 1905, Gilbert Grosvenor faced a critical decision—whether or not to publish photographs in the *National Geographic* magazine. Today, photography and *National Geographic* are synonymous. At that time, however, many intellectuals considered photographs superficial or even vulgar. Since its founding in 1888, the National Geographic Society had printed a dense, scientific journal featuring lengthy articles by the leading scientists, geographers, explorers, and intellectuals of the day. Grosvenor, the magazine's first full-time editor, was well aware that his decision could potentially anger the Society's Board and alienate its members.

But Alexander Graham Bell, the visionary scientist and former president of the National Geographic Society, had instilled in Grosvenor a strong belief that the relevance of the *National Geographic* magazine depended on publishing stories that were readable by 'ordinary people'. Otherwise, Bell said, they should 'shut up shop' and become a 'strict, technical scientific journal for high class geographers and geological experts'.[1] Grosvenor ultimately decided to print several

---

1   Alexander Graham Bell, 'Letter to Gilbert H. Grosvenor, 5 March 1900', Grosvenor Family Papers, Library of Congress, Washington, DC.

     https://doi.org/10.11647/OBP.0186.11

photographs of Lhasa, Tibet in the January 1905 issue. Even more radically, Grosvenor wrote captions that supported the photos, not the other way around, making the images the center of the story. His decision did indeed irritate several Board members. But the public loved the pictures just as Bell had predicted. Grosvenor continued to publish photographs, transforming the identity of the magazine and attracting new readers. In just two years, the Society's membership grew from 3,000 to 20,000. Today, photography is at the heart of who we are as a mission driven organization. When I tell people that I work for National Geographic, invariably the first question I'm asked is whether I am a photographer.

In 2018, on the 130th anniversary of our founding, the National Geographic Society recommitted itself to innovative storytelling. The Society launched a Media Innovation division that will fund year-long projects by several storytelling Fellows who are pushing the boundaries of what stories we tell — and how we tell them. The 2018 class included science photographer Anand Varma, environmental writer Emma Marris, digital storyteller Xaquín G. V., and Evgenia Arbugaeva, who grew up north of the Arctic Circle and whose project documents how people who inhabit the coastal stretch of land along the Northern Sea Route in Russia are adapting to political, economic, and climate changes.

The National Geographic Society has also revolutionized the museum experience by employing VR (virtual reality) technology in its popular Tomb of Christ exhibition, allowing museum visitors to more fully experience the Church of the Sepulchre. The Society also launched the first VR theater in Washington, DC, with programs that offer audiences more immersive experiences, such as diving with leopard seals in Antarctica and exploring Bears Ears National Monument in Utah. Wearing individual VR headsets, 450 people simultaneously experience what our explorers saw and heard in the field. This new technology has the potential to not only increase an audience's appreciation of what the speaker is presenting on stage, but also to develop greater emotional attachment to the places, people, and life that comprise the subjects of their talks.

In addition, in November 2018, we premiered National Geographic On Campus, a series of live science and storytelling events designed specifically for university audiences.[2] As program manager, my goal is to complement what students are learning in the classroom and to provide opportunities for students to amplify their impact by connecting them with the National Geographic Society's worldwide community of scientists, storytellers, educators, and explorers. These university-based events — open to all undergraduate and graduate students regardless of their field of study — center around two day-long programs: the Science and Storytelling Symposium; and a series of storytelling workshops led by National Geographic in areas such as conservation photography, investigative journalism, and transmedia storytelling.

The Science and Storytelling Symposium features dynamic talks and panel discussions that bring together National Geographic Explorers in conversation with university scholars and local thought leaders. The symposium highlights interdisciplinary thinking, science and storytelling collaborations, and the connections between research and storytelling. Program themes and panel topics center around regional issues and the university's disciplinary strengths and are developed in full partnership with each host university. For example, our pilot program at the University of Miami — World Without Borders — focused on sea level rise, freshwater scarcity, species conservation, wildlife trafficking, human migration, and cultural identity in an increasingly globalized world. Each symposium also features a panel called Storytellers for Change, which explores how storytellers are helping to create a more peaceful, just, and sustainable planet.

The second day features a series of National Geographic-led workshops that offer students an opportunity to learn and hone their storytelling skills with our Explorers and staff, including veteran Magazine reporters. It's one thing to be inspired by reading the *National Geographic*, or watching our films; it's another thing to see an Explorer live on stage, to hear the passion in their voice and to have the opportunity to ask them a question. But it's an incredibly unique opportunity for a student to spend a day working with a *National Geographic* photographer or journalist, gaining practical and conceptual storytelling skills that will enhance their work and impact regardless of their field of study. The goal is to both inspire and educate the students, and in the process,

2 'On Campus', *National Geographic*, https://www.nationalgeographic.org/on-campus/

cultivate informal mentor relationships for students in a variety of fields. All programs are offered free of charge to currently enrolled students.

The On Campus program is an outgrowth of the Science and Storytelling Symposium, a two-day event hosted by the Yale School of Forestry and Environmental Studies (F&ES) in partnership with the National Geographic Society in April 2016. As the curator of the symposium, my goal was to create the kind of experience that I wish had existed when I was a student — one that not only pushed me to think more deeply about storytelling, but also provided hands-on instruction and connections to professional storytellers. In addition to the daylong symposium that explored such topics as Artistic Representations of Nature, Meaning and Morality in a Contested Landscape, and Making Science Accessible through Storytelling, the event, conceived and developed with the support of Peter Crane, the Dean of F&ES at that time, featured a special presentation on Greater Yellowstone Migrations — a science-storytelling collaboration by ecologist Arthur Middleton, wildlife photojournalist Joe Riis, artist James Prosek, and filmmaker Jenny Nichols. This on-going project documents ungulate migrations in the Greater Yellowstone Ecosystem through a combination of quantitative science, cartography, and visual storytelling. We sponsored an art show featuring Prosek's and Riis's works, hosted an intimate fireside chat by science writer David Quammen, and held an exclusive dinner at the Peabody Museum of Natural History, with a special keynote address by Thomas E. Lovejoy. The National Geographic Society also hosted a workshop for their Young Explorers Grants program, which was open to students from throughout Yale.

The On Campus program seeks to replicate and upscale this student program. In addition to highlighting the work of *National Geographic* Explorers and Grantees, each live event will feature a range of scholars from the sciences, social sciences, arts, humanities, communications, business, and law. The National Geographic Society has always supported education; we develop curricula for middle school teachers and host a national geography bee competition, for example. And since our founding, the Society has sponsored live events, starting with the very first National Geographic Society lecture by John Wesley Powell in 1888. But On Campus is the first live National Geographic Society program specifically tailored to university audiences. Our inaugural event, held at the University of Miami, 9–10 November 2018, was followed by a second three-day student event at the University of Virginia in early March 2019, with a third planned for the University of Southern California. As we upscale the program, our hope is to partner with four schools a year (two each semester) — from liberal arts colleges and state universities, to historically black colleges and universities (HBCUs) and tech and engineering schools. We want to engage students from across the university in dialogues around science and storytelling, encourage them to think critically about interdisciplinary collaborations, and provide with them with storytelling tools to amplify their mark on the world.

In this twenty-first century, a degree is no guarantee of a job, or even of meaningful work. Students must not only develop deep theoretical understanding and demonstrated research in their chosen field, but also understand how to apply their knowledge and skills outside of the academy. Thomas Katsouleas, Provost at the University of Virginia, has referred to this as 'PhD Plus'. A program like On Campus can make a lasting impact on a student by marrying the disciplinary expertise they gained in the classroom with the storytelling skills, opportunities, and reach that the National Geographic Society can provide. Even the pioneering primatologist Jane Goodall has said that Louis Leakey made her a scientist, but National Geographic made her 'Jane'.

While On Campus is primarily intended for undergraduate and graduate students, a secondary goal of the program is to highlight alternative measures of impact for academic scholars. Despite the technological and cultural advancements in storytelling over the past century, many academics still dismiss photography and other storytelling media as illegitimate. Too often, scholars prioritize writing for an academic audience simply out of necessity because tenure is tied to publishing in peer-reviewed journals. Given the numerous demands placed on university professors — teaching, conducting research, writing, advising, giving lectures, and serving on committees, just to name a few — it's difficult at best (and, at worst, a liability) for scholars to spend time engaged in non-academic storytelling media, such as photography, film, podcasting, or digital storytelling. Publishing in scientific journals is undeniably a difficult and noble achievement, and engaging in meaningful dialogue with other academics can expand theoretical boundaries, generate research, and foster innovation and new ways of thinking. But scholars should also be allowed to pursue various storytelling media, and rewarded not only for the number of journal articles they publish, but also for the impact of their work beyond the academy.

I first began to appreciate the importance of storytelling while conducting a baseline presence survey of Canada lynx (*Lynx canadensis*), which had recently received federal protection under the Endangered Species Act, for the US Fish and Wildlife Service in Washington's Cascade Mountains. I had already conducted extensive field research on forest carnivores at Mt Rainier National Park as an undergraduate at the Evergreen State College, where I had learned to think of nature in terms of interdependent systems. Like other 'Greeners' (as we were known), I was required to attend weekly seminars where I read works of fiction and gained an appreciation for the role of the humanities in ecological thought. But although Evergreen is well known for its fine arts and documentary film programs, I did not pursue courses in storytelling that would have complimented my scientific skills. Moreover, I was not encouraged to develop creative ways to communicate my own scientific research; nor was I taught to conceive of scientific research as one story amidst a sea of stories about how the world works. Story, myth, legend — those were areas for the humanities, religious scholars, and philosophers. Film and photography — those were areas for artists and journalists. Science, on the other hand was empirical, objective, fact. Science, I had been taught, was not only the language of power when it came to the natural world, it was the correct worldview.

As a federal wildlife biologist, I was asked to attend a town hall meeting in southwestern Washington to discuss what the listing of the lynx meant for land owners and other local residents. I prepared my talking points and slides for a formal presentation. But upon arrival to the meeting, I was confronted by loggers, hunters, ranchers, and others who were angry that a small cat was threatening their livelihoods. As a federal employee, I was seen as the enemy. A veteran colleague who had anticipated such a reaction, however, arranged several round tables to promote dialogue between the local community and various stakeholders, including us government officials. I had come prepared to *talk* to the group about lynx. But, sitting there across the table from men and women who lived in this community, I found myself *listening* to their stories. I heard stories about their families, their grandparents, their neighbors. I heard stories about the woods, about their logging and hunting. I heard stories about how they had been down this road before with the spotted owl. I heard stories about how much the land meant to them.

As a field biologist, I had spent considerable time in those woods, but I didn't live there. I had never hunted, and I was a poor fisherman. I didn't know how to use a chainsaw. I didn't personally know any locals who actually lived there. In fact, I was born and raised on the domesticated prairie of north-central Iowa, not the mountains of western Washington. In short, I had no personal connections to the land beyond my field work and belief in the inherent value of the lynx. I had no real stake in the game, only data points and the confidence that comes with having science on your side.

Their stories made a deep impression on me. Not only did I not have a very compelling narrative, it was clear that their stories were about *them*, not lynx. Their stories were about their identity, which was deeply connected to the land. I wanted to hear more of their stories, and felt that somehow those stories were key to the protection of endangered species.

Some years later, as a graduate student at the University of Montana, I conducted an oral history project with the Confederated Salish and Kootenai Tribes who were at the time pursuing co-management of the National Bison Range, which is located in the heart of their reservation. Their efforts had generated a strong backlash by surrounding communities, including television commercials, newspaper ads, and town hall meetings where people angrily protested against co-management. I wanted to know why co-management was so important to the Tribes. I conducted dozens of interviews with tribal members, from the tribal chairman to elders to high school students. The Tribes had long accepted science as the language of power and developed a strong natural resource management program that included, for example, one of the first grizzly bear habitat management plans in the West. The stories I heard were once again about them as a confederation, about their history, about their deep cultural relationship with bison that went straight to the heart of their identity as a people.

When I later became a ninth-grade environmental science teacher, I came into contact with the actual beliefs of many young people, and with the ways in which popular culture uses stories to promote external values such as money, fame, and physical appearance as the ultimate measure of success. I saw how students gravitated toward those narratives, and identified the need to create curricula that encouraged students to not only think about science as a story, but also encouraged them to develop storytelling projects of their own. I incorporated art and roleplay into my climate change curriculum so that students would better understand different cultural perspectives and the complexity of global warming. Students produced short videos with their cell phones and created paintings that represented complex biogeochemical cycles. And during our field research, in addition to their data collection, I required students to keep a field journal of their qualitative observations that included both written reflections and drawings. I wanted my students to understand that science is more than a collection of data or even a systematic process of inquiry; it is a collection of stories about how the world works that says as much about who we are in this time and space as it does about that which we study.

The importance of storytelling is not a novel idea. Stories are how we make sense of the world, each other, and ourselves, and the world's greatest scientists are often powerful storytellers. But, in recent years, there has been greater attention to how we think about, create, and tell effective stories, particularly in the sciences, along with an explosion of new storytelling technologies. The climate movement, for example, has largely recognized the importance of storytelling to change people's hearts, minds, and behaviors, rebooting a fatalistic narrative to one of hope and action focused on people, not just polar bears. It is crucial for scholars and other researchers to consider the role that storytelling plays in their work, and to develop storytelling skills and interdisciplinary storytelling collaborations that engage, inspire, and educate the public not only about their research, but critically what it means for humanity.

Bibliography

Bell, Alexander Graham, 'Letter to Gilbert H. Grosvenor, 5 March 1900', Grosvenor Family Papers, Library of Congress, Washington, DC.

'On Campus', *National Geographic*, https://www.nationalgeographic.org/on-campus/

12. Listening for Coastal Futures

The Conservatory Project

Willis Jenkins

Eyes widened in surprise as each researcher put on the headphones. Even the two coastal ecologists who had been working here for decades had never before heard the sound. Just below the quiet water lapping at our feet, an oyster reef crackled. Tiny, snapping shrimp amidst the oysters make a punctuated vibration that sounds to human ears like a snapped finger or a dropped pebble. Collectively, these shrimp make it possible for us to hear the reef's structure, as a vertically layered crackling.

The designer of this listening station, the ecoacoustic composer Matthew Burtner, had placed one microphone atop the reef and another inside an oyster shell on the beach. In this way, we were able to hear two worlds simultaneously: the living din of an underwater oyster city, along with surface wind sibilating the dry carapace of a long-dead individual. Trying to still our feet from crunching the millions of shells on which we stood, we listened to the past and future of this coast.

***

The Coastal Futures Conservatory foregrounds listening as a form of inquiry that can be ecologically immersive and epistemically integrative. Working alongside a National Science Foundation (NSF)-funded Long-Term Ecological Research site at the Virginia Coast Reserve, the Conservatory connects scientific understandings of coastal change with modes of understanding cultivated in the arts and humanities. Its aim is to open the study of coastal change to multiple ways of knowing, and

   https://doi.org/10.11647/OBP.0186.12

then to interleave those knowledges in ways that stimulate integrative understandings of environmental change.

What does it mean to listen for coastal futures? The Conservatory organizes collaborative inquiry around listening in two basic ways. It is an embodied exercise in which all researchers participate, and also a guiding metaphor for attending to living worlds with multiple ways of knowing.

In the first, most literal way of listening, we convene cross-disciplinary research exchanges about coastal change within an intellectual space made by undertaking ecoacoustic listening exercises.[1] Focused attention with handheld microphones at a research site initiates the production of knowledge, by orchestrating an acknowledgment of our enfleshed immersion in soundscapes made by human and nonhuman vibrations. The exercise sets a contemplative threshold, and each researcher, whatever her discipline, comes into the seminar room having been compelled to meditate on her enfleshment in a soundscape made of human, nonhuman, and, especially, her own vibrations. The experience creates a platform for transdisciplinary discussion.

More importantly, the exercises begin to reorient our metaphors of environmental knowing from ocular to aural. North American environmental thought has often aspired to a visual model of knowing, captured in one of Ralph Waldo Emerson's most remembered lines: 'I become a transparent eye-ball; I am nothing; I see all'.[2] His vision suggests a disenfleshed beholding, as if a transparent spiritual presence receiving the self-manifestation of Nature. The Conservatory reorients that tradition by foregrounding the enfleshed ear — an organ that receives not 'all' but just some of those vibrations, mediating these, transforming them into somatic terms interpretable by our minds as 'listening'. Burtner pointed out that other forms of life would perceive the soundscape differently than we did. Each form of life hears the vibrations of the world differently, according to the constraints and affordances of its physiology.

E. O. Wilson, in *The Origins of Creativity*, writes that the limited sensory capacity of humans to perceive the activity of the living

---

1   The environmental recording work of Bernie Krause has been foundational for these kinds of exercises and for the emergence of acoustic conservation sciences.

2   Ralph Waldo Emerson, *Nature and Other Essays* (Mineola, NY: Dover Publications, 2009), p. 189.

world helps explain the anthropocentrism of work in the humanities. Epistemic anthropocentrism makes for cultural anthropocentrism, as meaning-making is bound by the affordances of human physiology. Incapable of knowing what it is like to perceive the world as a bat or an oyster, humans tell stories, sing songs, and create their grand projects without attending to the voices of other forms of life. Insensitivity to anthropogenic diminishment of Earth's life, suggests Wilson, stems from the cultural consequences of sensory limitation. The future of nonhuman life depends, in part, on the humanities learning to 'escape the bubble in which the unaided human sensory world remains unnecessarily trapped'.[3]

Our ecoacoustic listening exercises at once focus attention on the sensory constraints of human embodiment, while also expanding our capacity to hear. The apparatus of augmented aurality underscores the mediation of all perception. Listening with a microphone through headphones, my mammalian presence becomes perceptual foreground: the sound of my breathing, the brush of clothing with every small motion, the huge crunching and splashing sounds of a twelve-stone biped attempting to tread gently across a salt marsh. I do not become 'transparent', like Emerson; rather, I sense the life in this marsh — insofar as I do — as one enfleshed within it.

Yet, augmented aurality and designed listening stations also enhance capacities of attention. A hydrophone allows us to hear the crackling life of an oyster reef. That doesn't let us feel what it is like to be an oyster or a reef, but it does let us attend to oyster life in a different way. In that sense, focused listening cultivates a kind of empathy that can deepen scientific understanding, as Karen McGlathery, lead researcher of the Virginia Coast Reserve (VCR), observed. For the sake of honing research attention, McGlathery declared that every new researcher at the VCR should undertake listening exercises.

In fact, that particular exercise has already led to new research. When the environmental scientist Matthew Reidenbach heard the sound made by oysters feeding and filtering, he wondered if variations in oyster activity could be correlated with environmental conditions. With a pair of graduate students from Music and Environmental

---

3   E. O. Wilson, *The Origins of Creativity* (New York, NY: Liveright Publishing, 2017), p. 92.

Science, Reidenbach has deployed sensors to record oyster beds in various locations and conditions, in order to investigate whether there may be acoustic signatures of reef health and also to understand how oysters interact with their soundscape.[4] Among other things, it may turn out that anthropogenic vibrations from bridges and boat motors affect oyster activity, which would carry implications for the best places to locate aquaculture and reef restoration, as well as for the design of coastal development.

Designed listening stations push attention a step further by inviting listeners to connect what they are hearing with dynamics of ecological change. Connecting the hydrophone with a shore microphone drew listeners to connect the crackling now of oyster life with the untold cycles of life and death that created the shell beach on which we stood. Another station combined waves lapping against a mudflat with wind rustling through the sea grasses that hold the mudflat in place, inviting listeners to meditate on a micro-dynamic of sea level rise. Listening stations may perhaps cultivate empathic attention to patterns of coastal change.

Sonification of data takes that process yet another step. The Conservatory is transforming VCR data sets on sea level rise and wave action into something humans can hear, so that humans can listen to coastal change in still another way. Transforming data from visual to aural signals allows an audience to experience the science of environmental change in a different somatic register, which can resituate data reception within the immersive experience of hearing. Burtner's compositions then incorporate sonified data sets into musical expression and live performances, thus transforming the science of long-term ecological change into patterned vibrational changes that can be felt in one's body.[5]

Musical expression of long-term data of sea level rise allows audiences to reflect on the emerging species-level scale of human agency, as well as its inextricability from other dynamics of change. Informed

---

4   For more information on this research, as well as field recordings and sonifications of data, see www.coastalconservatory.org

5   'This music is built on techniques of environmental temporality and interrelated energy fluctuations, inspired by an uncommon way of listening to the natural world.' Matthew Burtner, 'Climate Change Music', *South Atlantic Quarterly*, 116.1 (2017), 145–61, at 146, https://doi.org/10.1215/00382876-3749392

audiences will know that that the rise in sea level is not entirely driven by anthropogenic forces, which rather interact with and intensify other forces in dynamic that operate over evolutionary time. Yet it remains difficulty to imagine. Thinking 'coupled human-environment systems' (also known as 'coupled human and natural systems') — to use NSF's clunky terminology — remains a grand challenge in part because responsibility for those systems often seems to depend on distinguishing human drivers within complex systems.

Where ideas of responsibility remain fixated on boundaries, the capacities of the ear may again afford an apt sensory basis for thinking into that imaginative challenge. The philosopher of sound David Dunn writes that 'when we look at the world our sense of vision emphasizes the distinct boundaries between phenomena. [...] In contrast, the sounds that things make are often not as distinct, and the experience of listening is often on perceiving the inseparability of phenomena'.[6] Burtner's compositions suppose that when we can hear a dynamic of environmental change, such as sea level rise, our minds are less likely to move to distinguishing and isolating; prioritizing aural over ocular experience disposes our minds to imagine ourselves immersed and participant within what we hear.

For Dunn, that makes music a critically important capacity for developing responsibility for human-influenced ecologies.

> It is a different way of thinking about the world, a way to remind ourselves of a prior wholeness when the mind of the forest was not something out there, separate in the world, but something of which we were an intrinsic part. Perhaps music is a conservation strategy for keeping something alive that we now need [...] a way of making sense of the world from which we might refashion our relationship to nonhuman living systems.[7]

If we need music to conserve worlds, then (say Burtner and I) we need a new kind of conservatory: a school of music practice that teaches participants how to listen to and compose with the living world, a school of science practices that conservation with culture, and a school

---

6   David Dunn, 'Nature, Sound Art, and the Sacred', in *The Book of Music and Nature: An Anthology of Sounds*, ed. by David Rothenberg and Martha Olvaeus (Middletown, CT: Wesleyan University Press, 2001), pp. 95–107, at p. 97.

7   Ibid.

of living that cultivates a wide range cultural capacities to listen to Earth and refashion our relationship with it.

As ecoacoustic composition with long-term ecological research (LTER) data sets creatively expands the experiential basis of our attention, it begins to accomplish something of what Wilson has called the humanities to do: to integrate with sciences in ways that expand sensory limitations of human embodiment and diminish the correlative anthropocentrism of value and regard. Hearing sea level rise within an affective performance makes it more likely that we will regard the phenomenon as significant for the stories by which we make sense of our lives, of our pasts and futures.

Those are interpretive questions for which the humanities are especially well suited. Two Conservatory humanists, the historian Andrew Kahrl and the literary scholar Charlotte Rogers, take cultural memory as their starting point — specifically, exploring how communities have already been listening to coastal change, telling stories about it, and even creating artistic performances in response. An historian of US coasts with a special focus on racial exclusion, Kahrl's research grounds 'coastal futures' in coastal pasts by taking oral histories of longtime Eastern Shore residents (another listening exercise) and by conducting archival research into shoreline development. Meanwhile, Rogers investigates the connected roles of hurricanes and slavery by following cultural responses to hurricanes along a geographic arc that traces the path of slave ships. Following 'the trace lines of trauma', Rogers finds a pattern of literary and artistic response to hurricanes that interprets their cultural meaning. Present in the work of both scholars is attentiveness to dynamics of race and power in shaping, especially along the eastern shores of North America, forms of coastal dwelling.

The Conservatory thus supposes that cultural capacities of environmental listening extend beyond literal listening exercises. In the work of Kahrl and Rogers, the Conservatory employs conventional humanistic methods in order to interpret the political and cultural frames within which Eastern Shore societies make sense of coastal change. In my own work, I am exploring listening as an ethical method and transdisciplinary practice fitting for questions of the Anthropocene.[8]

---

8   On the 'Anthropocene', see also 'Anthropology as Cosmic Diplomacy: Toward an Ecological Ethics for Times of Environmental Fragmentation' by Eduardo Kohn in this volume.

In religious and environmental ethics, we need scholarship that deftly pulls research toward grand questions: who is the human driving these changes? How should humans interpret and fit themselves into patterns of the world? What stories shall we tell of the past and future of the coasts?

Metaphors of listening are apt for organizing multi-layered research and inquiring into its significance with one another. Where visual models of integration might try to resolve multiple ways of knowing into a unified vision (which is Wilson's goal), an aural model of integration suggests hearing multiple different things at once. The Conservatory integrates arts, sciences, and humanities seeking resonance without resolution.

The literal ecoacoustic exercises nonetheless remain important. For our Conservatory researchers they offer a shared practice, a form of affective environmental inquiry that serves to ground our intellectual exchanges. With that shared aural experience, our conversation is primed to aspire not toward a unified vision, but toward a vibrating soundscape of thought. Moreover, Dunn holds that the highest possibilities of integration happen when listening is not only a metaphor of knowing but is also grounded in the actual somatic experience: 'the physical act of using our aural sense [...] can become a means to practice and engender integrative behavior'. For Dunn, the exercise of attentive ecoacoustic listening becomes a 'meditative practice' which can 'remind us of the sacred', leading us to rediscover 'our place in the biosphere's fabric of mind'.[9] Perhaps not all our researchers would put it that way, but participants in the Conservatory's first exercises did reflect that it put their inquiries in the context of something larger. Something about listening attentiveness to the living world can induce mythic wonder, which sometimes opens discipline-shaped minds to greater patterns of intelligence.

Being responsive to intelligences beyond the human may help us transform the impoverished notion of the human and the captive humanities. In 'The Human Shore' — a meditation on what the Anthropocene means for postcolonial Atlantic Studies — Ian Baucom writes that the 'interdisciplinary' era of intellectual life that began in tentative exchanges of the life sciences and humanities must now be

---

9   Dunn, 'Nature, Sound Art, and the Sacred', pp. 98–99.

deepened into something more transformative for the humanities — and for the human. Moving beyond the 'conflict of disciplines' toward a 'concert of disciplines' that collectively forges 'the era of a new humanism', Baucom observes the paradox that intellectual commitment to the deep future of the planet requires the humanities move beyond the human, drawing its critical ways of knowing beyond love of humanity itself.[10] Organizing a concert of disciplines within that paradox, it seems to me, points toward constellating arts, sciences, and humanities in practices of response to intelligences beyond humanity. Perhaps we can assemble the knowledges we need to reinhabit the human shore by listening to the still-living coast.

***

Can the disciplines learn to listen to Earth? In *Braiding Sweetgrass*, Robin Wall Kimmerer observes that while universities often invite students to change the world, we should instead summon them to listen to it. Where settler cultures are characterized by changing lands they have not bothered to understand, Kimmerer suggests that a key pedagogical task is teaching people how to listen to the land's stories.[11] Yet meanwhile, interdisciplinary institutions for sustainability or resilience invariably tell our publics that we will invent 'solutions' for problems of environmental change — implicitly promising them that we will find ways to keep Earth silent. Perhaps we should instead say that we do not yet have the necessary intelligence to imagine decent resolutions to the problems we are causing, but we are inventing ways to listen.

Listening to the land's story is not alternative to science. Kimmerer, an environmental scientist herself, describes the data-collection apparatus at the Andrews Forest (another LTER site), as instruments for 'listening to the land for stories that are simultaneously material and spiritual'. The land speaks a language we have either forgotten or have yet to learn. 'The archive of this language, the sacred text, is the land itself. In the woods there is a constant stream of data, lessons on how we might live, stories of

---

10   Ian Baucom, 'The Human Shore: Postcolonial Studies in an Age of Natural Science', *History of the Present*, 2.1 (2012), 1–23, at 9, https://doi.org/10.5406/historypresent.2.1.0001

11   Robin Wall Kimmerer, *Braiding Sweetgrass: Indigenous Wisdom, Scientific Knowledge, and the Teachings of Plants* (Minneapolis, MN: Milkweed Editions, 2013).

reciprocity, stories of connection.'[12] The question here is whether we listen to data as if they were elements of a language of living intelligence.

The imaginative task of listening to the land is vast for disciplines that have made the impossibility of doing so an axiomatic assumption of knowledge. Even when we work across arts, sciences, and humanities, our collaborations can enforce that epistemic anthropocentrism. Indeed, the model of 'integration' lauded by the National Academies of Science,[13] often means just adding knowledges about culture and people to knowledges about environments. Thus, Noel Castree observes, 'epistemic unity is vouchsafed by a presumptive ontological monism'.[14] Yet, the ontology of nature and culture in which Earth cannot speak and humans should not try to listen is — as Phillipe Descola has shown — a minority view.[15] In rejoinder to E. O. Wilson, it must be said that it is not all humans who experiences themselves as incapable of knowing other beings, and not all societies that have been so anthropocentric; those are modern North Atlantic accomplishments.

Recognizing the contingent historical character of the rule against listening to the land is especially critical in a time of global environmental changes driven by modern North Atlantic ways of life. For it is not just a matter of curiosity that other biocultural ontologies exist; we (who know in modern ways) seem to need capacities of listening to intelligences that our disciplines have excluded. At the end of his striking ethnographic account of *How Forests Think*, Eduardo Kohn writes: 'If "we" are to survive the Anthropocene [...] we will have to actively cultivate these ways of thinking with and like forests'.[16]

---

12 Robin Wall Kimmerer, 'Interview with a Watershed', in *Forest Under Story: Creative Inquiry in Old-Growth Forest*, ed. by Nathaniel Brodie, Charles Goodrich, and Frederick J. Swanson (Seattle, WA: University of Washington Press, 2016), pp. 41–49.

13 The National Academies of Sciences, Engineering, and Medicine, *The Integration of Humanities and Arts with Science, Engineering, and Medicine in Higher Education* (Washington, DC: The National Academies Press, 2018), https://doi.org/10.17226/24988

14 Noel Castree, 'Global Change Research and "the People Disciplines": Toward a New Dispensation', *South Atlantic Quarterly*, 116.1 (2017), 55–67, at 63, https://doi.org/10.1215/00382876-3749315

15 Phillippe Descola, *Beyond Nature and Culture*, trans. by Janet Lloyd (Chicago, IL: University of Chicago Press, 2013), https://doi.org/10.7208/chicago/9780226145006.001.0001

16 Eduardo Kohn, *How Forests Think: Toward an Anthropology beyond the Human* (Berkeley, CA: University of California Press, 2013), p. 227, https://doi.org/10.1525/

The idea of the Anthropocene seems for some an excuse to give up on listening. If ever Earth had something to tell us, goes one line of Anthropocene thought, the human roar now overwhelms it. Can we listen as beings whose mammalian embodiment is not only processing and mediating the incoming vibrations of the world, but is also creating reverb, intensifying its own vibrations and silencing those of others, entering into the chorus of Earth in dissonant ways?

Some of the old ways of practice cultivated in ancient traditions may offer analogies of practice for holding the paradoxes of nonanthropocentric knowing in an anthropogenic world. In several traditions of ancient practice, contemplative practice works to open oneself to the claims of the whole world by overcoming preoccupations with human cares, even while also holding that the world's existence depends, in some way, on human performance of ceremony or liturgy. They aim for nonanthropocentric existence in an anthropogenic world.

The scholar of Christian mysticism, Douglas Christie, has argued for 'an approach to ecology that understands the work of cultivating contemplative awareness as critical and necessary to its full meaning'. Ecological ways of knowing, he argues, need to be grounded in a sense of deep reciprocity with the world. We need 'practices that will help us feel and respond to the claim of the living world upon our lives'.[17] These may arise from recognized religious traditions, says Christie, but a contemplative ecology may also be created from other cultural sources. Christie combines premodern monastic practices with meditations from contemporary North American environmental thought. What the traditional practices offer are long-tested forms for dwelling in paradox and for overcoming human selfness. An ecological spirituality may take many different forms, writes Christie, but 'a common feature of such spirituality or spiritual practice is a deepening of awareness of oneself as existing within and responsible for the larger whole of the living world'.[18] Thinking about those practices amidst uncertainty,

---

california/9780520276109.001.0001. See also 'Anthropology as Cosmic Diplomacy: Toward an Ecological Ethics for Times of Environmental Fragmentation' by Eduardo Kohn in this volume.

17   Douglas Christie, *Blue Sapphire of the Mind: Notes for a Contemplative Ecology* (New York, NY: Oxford University Press, 2013), p. xi, https://doi.org/10.1093/acprof: oso/9780199812325.001.0001

18   Ibid., p. 20.

loss, and suffering in the context of so much negative environmental change, Christie writes:

> Here I believe is where spirituality, ethics, and politics converge. Without real feeling for the immensity in which we live and move and have our being, a sense of relationship and intimacy with it, will it really be possible to care for it? [...] It is here that the contemplative traditions of thought and practice, especially those that risk confronting the darkness, have so much to teach us. [...] It invites and perhaps even helps create in us a posture of humility: a willingness to listen, receive, and respond to all that is unfolding before us. But it also brings with us a great risk, inviting a relinquishment of self so profound it is in fact a kind of death.[19]

It may seem jarring, or at least intellectually indecorous, to frame the work of the Conservatory in terms of spirituality. Turning to religious arts of dying, however, indicates the depth of dislocation I think necessary to reconstellate knowledge practices for a world of rapid anthropogenic change. Not only do we need to reintegrate arts, sciences, and humanities; we need to develop skills of knowing that open human minds to listen and respond to other intelligences of the living world. As Castree puts it, we need the separated disciplines 'not merely to collaborate but *to unsettle each other* so that a new modus operandi emerges'.[20] The modes of *knowing* necessary for anthropogenic ecologies, I think, also reflect on themselves as proposed ways of *being* with the living world. I want to know of them: can they be transformed by what they hear?

On the Eastern Shore, as we listen for coastal futures, we assay to do so with all forms of coastal intelligence, human and nonhuman. As the oyster reefs interact with their soundscape, what are they telling us? As the barrier islands move with the seas, they embody a kind of intelligence we need for a world of sea level rise. What story are they telling?

---

19 Douglas Christie, 'The Night Office: Loss, Darkness, and the Practice of Solidarity', *Anglican Theological Review*, 99.2 (2017), 211–32, at 223–34.

20 Castree, 'Global Change Research', 65.

Bibliography

Baucom, Ian, 'The Human Shore: Postcolonial Studies in an Age of Natural Science', *History of the Present*, 2.1 (2012), 1–23, https://doi.org/10.5406/historypresent.2.1.0001

Burtner, Matthew, 'Climate Change Music', *South Atlantic Quarterly*, 116.1 (2017), 145–61, https://doi.org/10.1215/00382876-3749392

Castree, Noel, 'Global Change Research and "the People Disciplines": Toward a New Dispensation', *South Atlantic Quarterly*, 116.1 (2017), 55–67, https://doi.org/10.1215/00382876-3749315

Christie, Douglas, *Blue Sapphire of the Mind: Notes for a Contemplative Ecology* (New York, NY: Oxford University Press, 2013), https://doi.org/10.1093/acprof:oso/9780199812325.001.0001

— 'The Night Office: Loss, Darkness, and the Practice of Solidarity', *Anglican Theological Review*, 99.2 (2017), 211–32.

Descola, Phillippe, *Beyond Nature and Culture*, trans. by Janet Lloyd (Chicago, IL: University of Chicago Press, 2013), https://doi.org/10.7208/chicago/9780226145006.001.0001

Dunn, David, 'Nature, Sound Art, and the Sacred', in *The Book of Music and Nature: An Anthology of Sounds*, ed. by David Rothenberg and Martha Olvaeus (Middletown, CT: Wesleyan University Press, 2001), pp. 95–107.

Emerson, Ralph Waldo, *Nature and Other Essays* (Mineola, NY: Dover Publications, 2009).

Kimmerer, Robin Wall, *Braiding Sweetgrass: Indigenous Wisdom, Scientific Knowledge, and the Teachings of Plants* (Minneapolis, MN: Milkweed Editions, 2013).

— 'Interview with a Watershed', in *Forest Under Story: Creative Inquiry in Old-Growth Forest*, ed. by Nathaniel Brodie, Charles Goodrich, and Frederick J. Swanson (Seattle, WA: University of Washington Press, 2016), pp. 41–49.

Kohn, Eduardo, *How Forests Think: Toward an Anthropology beyond the Human* (Berkeley, CA: University of California Press, 2013), https://doi.org/10.1525/california/9780520276109.001.0001

The National Academies of Sciences, Engineering, and Medicine, *The Integration of Humanities and Arts with Science, Engineering, and Medicine in Higher Education* (Washington, DC: The National Academies Press, 2018), https://doi.org/10.17226/24988

Wilson, E. O., *The Origins of Creativity* (New York, NY: Liveright Publishing, 2017).

13. Imaginal Ecology

Brooke Williams

Looking back over my adult life, most of it has been about wildness — wandering in wild places, or wondering how to save them.

Years ago, I came to the conclusion that we need to save wild places because they save us. According to Henry David Thoreau, 'in Wildness is the Preservation of the world';[1] I spend most of my time thinking about why this is true. Contemplating this has led me to Carl Jung and the collective unconscious. I read everything I could find on the subject, including a book called *The Earth Has a Soul: The Nature Writing of Carl Jung*, edited by Meredith Sabini, a psychologist working in Berkeley, California.[2] She was very kind to me and tried to answer my questions.

Through her, I came to better understand the collective unconscious as a container for the entire evolutionary history of our species, including all we've ever needed to save ourselves. Natural history is key to understanding how our exposure to the wild world provides access to the collective unconscious. Those who choose to study this are, therefore, 'Imaginal Ecologists'.

One day, we were talking on the phone. She said, 'my colleagues and I have been talking about you.'

'Hmmm,' I said. 'Why?'

'Our sense is that whenever our species has been in trouble, the collective unconscious rises to the surface to help us, in often surprising ways.'

---

1   Henry David Thoreau, 'Walking', *The Atlantic Monthly*, 9.56 (1862), 657–74, available to read online at https://www.theatlantic.com/magazine/archive/1862/06/walking/304674/

2   Carl Gustav Jung, *The Earth Has a Soul: The Nature Writing of Carl Jung*, ed. by Meredith Sabini (Berkeley, CA: North Atlantic Books, 2002).

   https://doi.org/10.11647/OBP.0186.13

'Yes?' I said.

'When someone as naïve as you starts asking so many questions, we must really be in trouble.'

I took that as a compliment.

Part of a book I'm currently working on is about dragonflies, based on fifteen years of the notes I've taken recording my encounters with dragonflies. These encounters began with a dream I had involving the image of a dragonfly.

My chief interest in dragonflies, beyond the fact that they're such amazing biological creatures, is that, since they are born in water where they may live as nymphs for years before emerging as adults, symbolically, they are seen as the messengers between worlds, bringing contents from our inner unconscious into consciousness. Water, to renowned psychologist, Carl Jung, 'is the commonest symbol for the unconscious'.[3]

I had planned to recount more dragonfly stories for this anthology, but changed my mind. Instead, I would like to invite you to participate in an exercise for invoking imaginal ecology that I often do with my students.

***

So. Please stand with your knees slightly bent. Breathe deeply a few times... three counts in, hold it for three counts. Then exhale slowly for four counts. Then hold for four more.

(Thirty seconds pass).

You are in a dark basement. This basement is so dark you cannot see your hand that is waving in front of your face. You stand motionlessly, not wanting to bump into anything. (A few seconds pass). Your eyes adjust to the darkness and you are able to differentiate shapes around you. And then, in the distance, across a large room, you see slivers of light entering through the cracks forming around what seems to be door. You walk toward the door.

You feel the surface of the door, hoping to find a knob, which you do, midway down near the right side.

Turning the knob frees the door and pulling it toward you floods the room with bright sunlight.

---

3   Carl Gustav Jung, *The Archetypes and the Collective Unconscious* (Princeton, NJ: Princeton University Press, 1981), p. 18.

The sunlight is so bright that you shield your eyes from the brilliant and burning whiteness.

Your eyes adjust and you realize that you are standing at the bottom of a staircase. Describe that staircase. What is it made of? How many steps can you see?

Stepping up onto the first step, you look down and notice your foot? Is this actually 'your' foot? Is it covered? If so, by what?

You walk up the steps.

From the top step you see, spread out in front of you, a familiar landscape. Describe it.

Looking down, you see that you stand at the beginning of a path. Describe this path.

You step out and begin walking along this path.

In the distance you see someone or something moving toward you on the same path.

As you move closer to this other being, you see that it is an ancestor. Who or what is this ancestor?

You meet.

Your ancestor has a gift for you. What is this gift?

You thank your ancestor for the gift just given, turn, and walk back along the path the way you have come.

Where the path ends, you step onto the stairs and walk down. You enter the basement through the same door, shutting it behind you. Once again you find yourself in the dark room.

The light comes on.

***

The conference had a different feel from others I'd attended. The setting — with its beautiful art and gardens, the 'basket room' where our sessions were held — became for me the perfect container. The people, so open and ready, made me feel that I was part of some strange and wonderful family. I felt a nearly immediate bond with everyone, and had the strong sense that everyone there was looking out into the world wondering how, or if, their work contributed to a better understanding of how we might make it through this dangerous moment in earth's history, due mainly to the impact climate change is having on all life, and the US Government's refusal to acknowledge it.

After my presentation, many participants approached me to tell me what had happened to them during the exercise: mainly describing who their 'ancestor' was, and the gift they'd been given. While I hold those conversations in confidence, I'm comfortable describing my experience. I'll begin with the exact notes I made that day.

The stairway. How weird. It was paved, like an old highway with a double yellow line up the middle. Crumbling a bit at the edges. Abandoned, re-routed for efficiency, perhaps. A symbol of the two-lane roads I've always said were important in the desert to get between places of importance.

My foot/the shoe, a strange-colored Chuck Taylor autographed high top. At one time in my life, I wore them for everything — basketball, but also squash. I ran in them and wore them while hiking in the desert. I always waited for them to go on sale when I would buy three of four pairs.

The landscape, looking west on Highway 50–6 between Green River and Wellington, Utah, the two distant perfect buttes, I've only recently discovered.

The path, a recently graded dirt road through the very dry desert, wide enough for a large vehicle. Lined by low-lying, yellowing plants. Curving to the left.

The ancestor, an older, unknown bright-eyed woman — someone who looked like Julianne Warren might a few decades from now. She wore a flowing brightly colored dress or coat.

The gift. She leaned in and whispered into my right ear: '*My gift to you is perpetual spring*'. I sensed that she was not referring to the season but rather the feeling of moving from the dark and cold into the light and new life. The hope that this brings. She may as well have said 'perpetual hope'...

I am thinking back on this experience now, nearly four months later. Once again, I'm astounded by the degree to which every moment of life, if considered, if *attended* to, is filled with symbols such as these. Because of the temporal distance between me and this experience, I now find myself comfortable not trying to recreate what these symbols meant back in October. These symbols are 'active' in that they're alive; they will 'shift' the story to one I need now.

I like that idea of *attending* to life, of *paying attention* to it. Where does this concept of *paying* attention come from... as if buying something? No, as if giving up something to acquire something else? Is this what *paying* attention is?

So, I return first to the **stairway**. I wrote recently that nothing says post-apocalyptic like a crumbling highway. This is what comes to me today. That the stairway in my 'dream' was a crumbling highway suggests that it would soon disappear; it hadn't, because of some massive shift in priorities, been maintained. And where once it was a well-used 'way' from one place to the next, it had long ago been forgotten, and had thus fallen into disrepair. The 'way' between the lower dark world and the bright upper world, once obvious, well-known and constantly used, had been abandoned. The 'way' still exists. My job is to repair and maintain it and encourage people to begin using it once again.

The **shoe**. My memory suggests that the Chuck Taylor All Stars I wore in my 'dream' weren't so old as they were intended to look old. Fashionable, in the way things like distressed jeans are these days. Something 'old' and forgotten has come back to be popular. I noted that during the sessions in the library on the last day of, Mitch Thomashow, one of the conference participants, wore grey, modern, Chuck Taylors to make his presentation. Two things occur to me. First, that I was wearing shoes that were once popular and very useful and have recently become fashionable again, may coincide with the stairway — *my job is not only to repair and maintain the 'way' between worlds, but also to personally move quickly and precisely along it*; after all, Chuck Taylor's are still athletic shoes. Second: wearing someone's shoes could mean seeing that person as an example for my future. That these shoes were once mine and now Mitch's fascinates me. Mitch and I are about the same age. Mitch has been on an amazing path as an educator, writer, wanderer, and wonderer. That I wore these specific shoes in this 'dream' suggests that I might find out more about Mitch, that I might discover something in/ about him that *resonates* for me as I move forward.

('Resonate', as I use it here, is a feeling that an idea or image 'fits' into place. I also like the term 'register', which is a printing term meaning 'the exact correspondence of position of color plates'. The best way I can describe it is a roulette wheel with all the holes representing possibilities. My inquiry is the silver ball that bounces among all possible answers

Fig. 7 Unnamed Buttes South of Wellington, Utah. Photo by Terry Tempest Williams (2019).

before falling into the perfect hole. The sound the ball makes as it drops into the hole — that is what I mean by 'register' or 'resonate').

The **landscape**. I've driven that highway section fifty times but never noticed those perfect buttes until last October, driving north from our home in Castle Valley to fly out of Salt Lake City for the conference. I looked for the buttes the next two trips but didn't see them. Finally, driving back home after the holidays, moving slowly, we saw them — the bright sunlight faded the background, setting them apart in contrast. Barely higher than their surroundings, they seemed vaguely like the twin buttes for which Bears Ears National Monument is named. My sense is that these 'new' buttes are in some capacity my own personal 'Bears Ears'. While I've driven by the actual buttes referred to as 'Bears Ears' for years, until we began working on National Monument status for the area surrounding them, I had no idea of their role in stories from the deepest past. The Bears Ears buttes represent the 'sacralizing' of that place. Of all the land conservation issues I've been involved in over the past four decades, Bears Ears may be the most important, because of the emphasis on sacred lands. I believe that this issue is intended to show (or, indeed, remind) us Anglos the importance of 'sacred lands'. According to the native peoples involved in this monument designation, 'It is time'. These new twin buttes demand my attention. I'm making plans to walk out to them during my next trip. My wife, Terry, took a photo (see Fig. 7).

The path. The dirt road in my 'dream' was much like many of those I documented during my time as a 'Field Advocate' for the Southern Utah Wilderness Alliance, a decade ago. Thousands of miles of these roads crisscross western deserts, most made in early efforts to exploit the area's resources. My job was to 'ground truth' maps — to identify which of the 'roads' shown on the maps still exist on the ground. In my dream, the 'road' (my path) has been recently 'improved', as it is lined on each side by spill from the grader. The degraded condition of the plants indicates late summer drought. Symbolically, I use this road to access desert resources. What resources am I looking for? I realize that lately I've been wondering about *desert wisdom*, and about how early seeds for Islam, Judaism, and Catholicism were planted by the 'Desert Fathers'. I'm searching for an understanding of what *exactly* that 'wisdom' comprises. It, once defined, may be the most valuable desert resource yet.

My **ancestor**. My recently graded road/path allowed me access to my ancestor, a beautiful mature woman — how I imagine Julianne Warren in a few decades. And if the road had really been graded to enable the extraction of the 'desert wisdom' resource, then this ancestor must symbolize the embodiment of that resource. Recalling my vision from October, I sensed her creativity based on her confidence and colorful clothes. The positive energy emanating from her attracted me. Today, I see this woman as my muse and wonder why my muse appeared in my 'dream', and not an ancestor. But then I catch myself and recall what I've learned about our dead ancestors — that they may be out there all around us, knowing much that we don't know, and wanting to help us. A muse is a person, usually a woman, who is a source of artistic inspiration. A dead ancestor wanting to help could easily come in the form of a muse.

'*My* **gift** *to you is perpetual spring*'. My first thought is to wonder if, as I thought last October, my ancestor — my muse — referred to the season 'spring'. Or '*a* spring', as in a continual font of pure water bubbling up through the earth. Then, I realized, that she did not whisper 'a perpetual spring', but rather 'perpetual spring' — without the indefinite article — as in 'springtime'. Once I sorted this out, I focused on the feeling I get during the first specific glimmers of spring — the way the light changes and the smell of that moment when I turn the first water onto the trees we've planted. And the different warmth coming from the sun. The sounds of the first meadowlarks singing from the tops of the sage. Today, writing this in Cambridge, Massachusetts, blanketed in last night's fresh snow, these images of spring do not send me longing for the future, but imbue me with a strong and vivid sense of possibility and positivity. If, like the months, I keep moving forward into the future.

***

I use this exercise in the college courses that I teach. Often, different elements generated by it become significant elements in the stories the students write. I always tell them that I'm not sure where these images come from. I'm not sure if truly paying attention has any evolutionary value, but I'm also not sure that it doesn't. 'I know one thing', I tell them, ' — that we don't seem to have solutions for all the problems we're creating. We need to learn how to dream new solutions.' I don't know that this exercise has anything to do with that. But I don't know that it doesn't.

Bibliography

Jung, Carl Gustav, *The Archetypes and the Collective Unconscious* (Princeton, NJ: Princeton University Press, 1981).

— *The Earth Has a Soul: The Nature Writing of Carl Jung*, ed. by Meredith Sabini (Berkeley, CA: North Atlantic Books, 2002).

Thoreau, Henry David, 'Walking', *The Atlantic Monthly*, 9.56 (1862), 657–74.

SECTION V

RELATIONSHIPS OF RESILIENCE
WITHIN INDIGENOUS LANDS

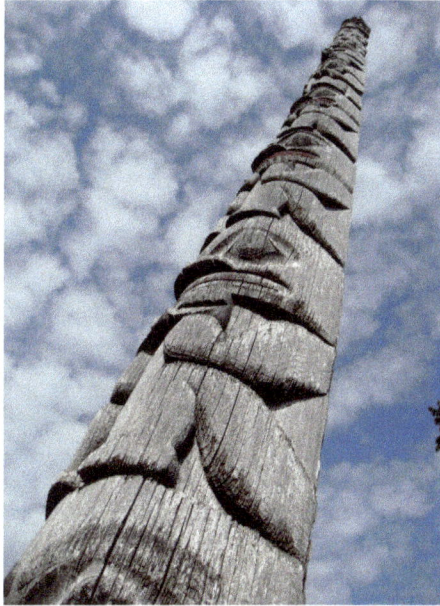

Fig. 8 Totemic Sentry, Prince Rupert, British Columbia, Canada. Photo by Jason Hall (2008), CC BY.

14. An Okanagan Worldview of Society

Jeannette Armstrong

I grew up in a very remote part of the Okanagan on the Penticton Indian reservation in British Columbia, Canada. I was born at home on the reservation, and I was fortunate to be born into a family that was considered by many people in our area as a traditionalist family. I grew up in a family where the first language was Okanagan, and which practiced hunting/gathering traditions on the land. I'm still immersed in that family. I've lived that life and I continue that practice in my own family. Growing up in a community that was small and fractionalized — fractionalized both by colonization, and, in many ways, in terms of the community itself — has given me valuable insights and observations. I thus have two perspectives from which to look at society — the perspective and experience of my small extended, traditional family support system; and the perspective and experience of a community that has been fractionalized by colonization.

One of the primary observations I wish to make centers on human relationships — the relationships that we have with one another, and the way in which these relationships impact our interactions with the land. Some, indeed many, of the changes experienced by our land have to do with our relationships with other humans — what we do to each other, and how we look at each other. In order to understand our relationship with the land, we must look closely at the relationship that we have with one another. I grew up in a community, in an extended family. In this community, people organized themselves in a very different way than that which I have observed outside of our

 https://doi.org/10.11647/OBP.0186.14

community. I want to describe how this community organizes itself, and to outline my perspective to you.

The land that I come from is similar, in terms of climate, to California. It is very dry and semiarid. It is considered the northern tip of the Great Basin Desert, and its ecosystem there is very, very fragile. Indeed, the Okanagan is one of the most damaged areas and ecosystems in Canada, due to its fragility. We live in an area where there are many conservationists and environmentalists concerned about the endangered species in the Okanagan, and where extirpations have been occurring over the last hundred years. I have seen some of those extirpations firsthand.

This experience has been difficult for our community, because we grew up loving the land. We grew up loving each other on the land, and loving each plant and each species the way we love our brothers and sisters. This form of love is not a result of an intellectual process. It is not a result of needing to gather food and needing to sustain your bodies for health. Rather, it is a result of how we interact with each other in our families, in our family units, in our extended family units, and in our communities. It is a result of the networks that we develop outwards from our families, extending to the other people surrounding us in our community. These networks are an essential part of how we interact with the land. My work hinges on interpreting these networks, and on reconciling members of my community, in order to restore health to the land. I am only able to do this responsibly if I have been able to generate a sense of understanding. In the Okanagan, our understanding of the land is one in which we are not just *part* of the land, nor just *part* of the vast system that operates on the land, but that *the land is us*. In our language, the word for our bodies contains the word for land. Our word for body literally means 'the capacity for land-dreaming' — the first part of the word invokes my ability to think and dream, and the latter part of the word invokes the land. Therefore, every time I say the word for my body, I am reminded that I am from the land. I'm saying that I'm from the land and that my body is the land.

Our community loves to go out to the land to gather, which I have continued to do every year of my life, and which I look forward to every year. I go out to the land to gather the foods that have given me life and given my grandmothers, and my great-great-grandmothers, life for many generations. We have perfected a way of interacting with each other when we do so, that is at once respectful to the land and respectful to each other. Our grandparents told us that the land feeds us, but that we feed the land as well. What they meant was that, in our very language, we give our bodies back to the land physically. In turn, we live on the land and we use the land; we can impact the land and we can destroy the land. Or we can love the land and it can love us back.

So, one of the things that I examined in the development of our education program at En'owkin Centre was how to teach about the way in which we, as a society, interact. My aim was to explore how our community interacted with each other, and to find a way to distill, describe, and teach this form of interaction, so that we might reconstruct it in our communities.

In doing so, I started to understand that the way in which we make decisions, and in which we choose to look at each other as people — as equal human beings — is fundamental to how we interact with the land. In the most basic sense, our use of the land relates to our need for food, for shelter, for clothing, and so on. When we look at society, we need to look at how society is constructed. Those are the things that we need. Those are the things that we need in order to live and breathe every day. But besides these basic essentials, we need pleasure. We need to be loved, we need to have the support of our community, and the love and the care of the people surrounding us. If we consider how these two necessities (our need for food and shelter; and our need for pleasure) are connected, and how they work together, then we can begin to understand how we might impact the land either in a negative way or in a positive way.

When we observe how the land has been impacted by western culture, we see that there is, at once, an overuse of resources, and a lack of access for some people to these resources. In other words, some people have more of a right to resources, and some people have less, or, indeed, no right to these very same resources. Within this system, there are also people who cannot access the basic things that they need to live. There is something profoundly incompatible between my idea of democracy, and the reality of a hierarchical system in which people, living side-by-side, do not have the same access to resources as one another. Equal access is a profoundly basic principle in our community — equal access to food and shelter; and equal access to pleasure and enjoyment of life.

I therefore started looking at decision-making as a construct. I looked at the traditional, historical Okanagan decision-making process, elements of which are still present in our community today, and have been carried forward because we're only two generations along from our colonizations. In our traditional decision-making process, we have a word — *enowkinwixw* — based in an image of people helping each other to absorb information like droplets of rain. This word demands four things from us: that we solicit the most opposing views; that we seek to understand those views using non-adversarial protocols; that we each agree to be willing to make adjustments in our own interests to accommodate diverse needs expressed; and that we collaboratively commit to support the outcomes. These are the four things that constitute an informal process that is continuously at play in our community. We can also engage with the process in a more formal way — in which case, it is known as the Four Societies process. When this is done, it functions as a construct, in the same way as, for example, Robert's Rules of Order, or the modern construct of democracy functions. However, the basic democratic construct entails that the majority has decision-making power over the minority. From my perspective, an adversarial approach is embedded in this construct. It sets up the oppression of the minority and establishes conflict at the heart of the construct, since there will *always* be people in the minority and people in the majority. It engenders dissension. I understand that it is an easy and practical approach to decision-making. But, in terms of the outcome this decision-making process produces — for society and for the land on both a local and global level — it seems to me that we must systemically rethink it.

From our point of view, the minority voice is the most important voice to consider. It is the minority voice that expresses the things that are going wrong, the things that we're not looking after, the things that we're not doing, the things that we're not being responsible toward, the things that we're being aggressive about or overlooking. One of the things our leaders said in the Four Societies process is that if you ignore this minority voice then it will create conflict in your community, and this conflict will create a breakdown that endangers the *whole* community. This conflict will endanger how we cooperate, how we use community as a process, how we think of ourselves as a cooperative unit, a harmonious unit, a unit that knows how to work together, that enjoys working and being together, and that loves one another. If such a breakdown occurs, then it starts to affect those things we need to do every day in order to meet all of our needs. I can see this in action in our world today. If we begin to think about the minority, about *why* there is a minority, why there is poverty, then we should be able to find creative ways to meet the needs of the minorities. Is it about economics? Is it about societal access? Human creativity is capable of identifying how to meet the needs of those minorities. It will enable us to bring that minority group into balance with the rest of the majority. This process is what we call *enowkinwixw*. If we are unable to enact this process in our community, then our humanity is at stake, our intelligence is at stake; we can't call ourselves Okanagan if we are unable to provide for the weak, the sick, the hungry, the elderly, and the disadvantaged.

In the same way, one component of our decision-making process is reserved for the land. This component involves individuals known as land speakers. I have been fortunate to be trained and brought up as a land speaker in my community. Unlike other communities, our community has people who are trained as part of a family system to be a speaker for the children, to be a speaker for the mothers, to be a speaker for the elders, to be a speaker for the medicine people, a specialist group of helper practitioners, to be a speaker for the land, to be a speaker for the water, to be a speaker for all of these different components that make up our existence. As land speaker, I have been trained by elders to think about the land, to speak about the land. This does not mean that I necessarily represent their view, and I do not consider myself an expert; rather, my constant responsibility to my community, no matter the decision in question, is to stand up and inform the community on how that decision is going to impact the land. How is it going to impact our food? How is it going to impact our water? How is it going to impact my children, my grandchildren, my great-grandchildren? What will the land look like at that time? The Four Societies process, *enowkinwixw*, is therefore founded on this principle of human interaction.

Another part of the process requires people to look at relationships. How is this decision going to impact the children? What are the children's needs? What are the elders' needs? What are the mother's needs? What are the working people's needs? It is someone's responsibility to ask these questions. When this person asks these questions, they must also give their views on the situation, just as part of our community is asked to think about the actions that need to be taken.

Part of our community stands up and says, what are the things that need to be built? What are the things that need to be implemented and how much is it going to cost? And all of those important action details need to be asked about and discussed. Those people in that part of our community who are speakers and doers are given the responsibility of continuously reminding our people that actions are going to have impacts — both short-term and long-term. If we overuse a resource, there are people whose role it is to stand up and inform the community of this.

An additional group of people in our community are known as the visionaries, the creative people. These are the artists, the writers, and the performers whose responsibility it is to bring innovative perspectives into the community, informing us of creative approaches, and new ways to look at things. These visionaries remind us that we must always make room for innovation, and that creativity is necessary in order to resolve issues that we haven't faced before.

In this way, all four of these components participate together in a decision-making process. Through this collaboration, the process then becomes a different process than Robert's Rules, or indeed the modern construct of democracy. The process becomes something that is participatory, that is inclusive, and that gives people a deeper understanding of the variety of components that are required to create harmony in a community. By incorporating the perspective of the land in terms of human relationships, the community changes. People in the community change.

Something internal happens, where people begin to realize that material things themselves are meaningless; that it is not material wealth that secures and sustains you and protects you from fear. Rather, it is the people and the community that secure and sustain you. When you become immersed in this belief, all fear leaves you, and, instead, you are imbued with hope — the hope that others around you in your community can provide these things.

This is the kind of work for the community that I'm involved in at the En'owkin Centre, which is a non-profit cultural educational organization governed by the seven reservation communities of the Okanagan Nation. I'm talking about all of the community. I'm talking about all of the people who live in the Okanagan and people that we reach outside of that. Not just the Indigenous people, because at this time in our lives, what our elders have said is that unless we can Okanaganize those people in their thinking, we're all in danger in the Okanagan. While it sounds simple, it often seems an overwhelming task.

Some days, it seems as though one person cannot make any difference. But, I think about my aunt, who spoke to me the other day, and asked me, 'Where are you headed off to now?' I replied, 'Oh, I'm going to this conference, the Bioneers Conference.' And she asked, 'Oh, what is that about?' So I did my best in my language to explain it to her that it is a conference that describes its purpose as bringing forward breakthrough solutions for people and the planet. And she said that kind of conference is a really good thing. She said, 'How did you come to be asked to speak at that conference?' And I said, 'I'm not really sure, but I think I managed to do that by talking and writing about some of the things that seem everyday and simple to us. That seem to make sense to us, that seem to transform complete strangers into our loved ones, by bringing them into our community so that they become part of my family and part of my extended community'. People like Fritjof Capra and Zenobia Barlow and other individuals who are friends of this community, and part of this movement. They feel the same as my aunt to me. I think that this is how we need to relate to each other. In doing so, we will begin to understand how we relate to the land, and we can begin to liberate ourselves from our perceived dependence on material things — such objects that make us feel secure and empowered, that tell us, 'you need a new car, you need lots of money, etc.'. Once this dependence begins to dissipate, then we begin to understand that the power is *us*. That *we* are our security on the land. And that that's what's going to sustain us.

The last thing that I want to share with you is my father's observation on insanity. For us, one meaning of insanity is too many people talking simultaneously about different things, as opposed to people collectively talking about the same thing. There seems to be this kind of insanity in the world because of something currently missing in terms of how we conduct our humanity with one another. When we start to address our relationships with one another, the land, in turn, has an effect on us.

One final observation I leave you with, is the power of taking our young people out to the land, to participate in the work we do to gather seeds and other Indigenous foods. Our community has started a program to replant the habitat needed for some of the endangered species that provide Indigenous foods. In order to replant Indigenous plants and to restore endangered habitats in the land that En'owkin is caretaking, we are growing about 10,000 plants each year. We have done so both to sustain ourselves and to sustain these species. The process of being with people, collaboratively, on the land, is a fundamentally healing process. This practice of gathering, potting, and replanting seeds to provide this habitat has proved immensely popular among all sorts of individuals from the non-native community, from multicultural societies, and from elderly communities. It is an especially valuable, healing process for struggling young people. The value of this process resides not simply in collecting seeds — but in being with people, the community, and communing with each other. It is how the land communes its spirit to you, heals people, and it does this in an incredibly profound way. We need to think about how we can do more of this.

15. Indigenous Language Resurgence and the Living Earth Community

Mark Turin[1]

Endangered languages and the communities that speak them are under extreme stress. Even conservative estimates paint a picture of near-catastrophic endangerment levels and possible collapse, with half of the world's remaining speech forms ceasing to be used as everyday vernaculars by the end of the twenty-first century.[2] The pressures facing endangered languages are as severe as those recorded by conservation biologists for flora and fauna, and in many cases more acute.[3] Yet linguistic endangerment is by no means a natural or inevitable process, the unfortunate by-product of modernization. Rather, the marginalization and erosion of local and Indigenous languages is the direct result of colonization and the racist policies that accompanied it. Across the world

---

1   This contribution has benefitted greatly from generous feedback from fellow participants at the original workshop that brought us together, in particular, Jeannette Armstrong, Sam Mickey, Mary Evelyn Tucker, and John Grim. In addition, I am grateful to Aidan Pine for the many deep conversations we have shared and collaborative writing projects that have helped to refine the points contained in this contribution, in particular: Aidan Pine and Mark Turin, 'Language Revitalization', in *Oxford Research Encyclopedia of Linguistics*, ed. by Mark Aronoff (New York, NY: Oxford University Press, 2017), https://doi.org/10.1093/acrefore/9780199384655.013.8
2   Michael Krauss, 'The World's Languages in Crisis', *Language*, 68.1 (1992), 4–10, https://doi.org/10.1353/lan.1992.0075
3   William James Sutherland, 'Parallel Extinction Risk and Global Distribution of Languages and Species', *Nature*, 423.6937 (2003), 276–79, https://doi.org/10.1038/nature01607

---

and through a variety of efforts that have included education initiatives, punitive legislation, and intentional neglect, colonial authorities have instituted language policies that sought to weaken traditional cultural practices, assimilate Indigenous populations, and gain access to their land and resources.

Colonial authorities have used the power of language and the language of power to further their own strategic ends. In some cases, and seemingly paradoxically, this involved supporting Indigenous languages; in most cases, however, they sought to erode them. In the first instance, believing in the inherent superiority of Christian theology, many missionary linguists focused on translating scripture into Indigenous languages. In Papua New Guinea and other regions of the Asia Pacific, scholars and administrators actively strengthened Indigenous languages through standardization programs that involved grammatical descriptions and the compilation of dictionaries and other pedagogical tools.[4] The goal — in many cases — was for local languages to be harnessed to transmit and disseminate an imagined Christian modernity. In other instances, as in Canada, settler-colonial authorities observed the unique relationship that existed between a language and the land on which it was spoken, and focused their attention on breaking this relationship apart by destroying the language and forcibly relocating communities far away from their traditional territories.

To this day, Indigenous communities around the world make use of traditional place names to ascribe current or historical meaning to places and spaces that are locally resonant and historically important. These powerful toponyms encode lived experience and traditional ecological knowledge in an ancestral language in a way that is almost impossible to translate into a more dominant national or international language. By disconnecting the language traditionally used to refer to a specific site, and by introducing new place names in a colonial language (the terms 'New Zealand' and 'British Columbia' serve as enduring examples), the relationship that local peoples have with their land was rendered opaque and further attenuated. Having weakened this connection to

---

4   S. Wurm, P. Muehlheausler, and D. Laycock, 'Language Planning and Engineering in Papua New Guinea', *New Guinea Area Language and Language Study*, 3 (1977), 1157–77.

land, the colonial goal of relocating communities in order to extract resources from their territories became more achievable.

Yet, for as long as efforts have existed to impose colonial languages on Indigenous peoples as a means of reshaping their identity, these same processes have been vigorously opposed by speakers of these languages. Pushing back against the decoupling of language from landscape, and asserting the uninterrupted continuity of a living earth whose community is sustained and nurtured by the intergenerational transmission of traditional cultural knowledge, Indigenous peoples find themselves at the front lines of environmental struggles that intersect with de-colonial forms of political activism. Opposition to externally imposed language policy takes many forms, from active resistance to passive non-compliance. Everyday forms of resistance have included the direct avoidance of colonial education programs by concealing children and evading census enumerators, to more contemporary and structured efforts in support of language revitalization, reclamation, and the renaming of traditional territories.

The emergence of the Caribbean linguistic mosaic can be seen as an anti-colonial response predicated on 'the need to speak and not be understood by the downpressors (slave masters, elite of society)'.[5] Viewed in this light, the creation of a *patois/patwah* or *creole/kweyol* can be read as a linguistic manifestation of a moral objection to the imposition of a hegemonic identity advanced by an imperial state, a perspective further substantiated by the Métis of Canada, who 'moulded the aboriginal and settler languages into coherent patterns which reflected their own cultural and historical circumstances'.[6]

Universities and municipalities in Canada are increasingly introducing statements that acknowledge Indigenous lands, treaties, and peoples, and also engaging in highly visible renaming practices that replace colonial-era names of buildings and places (usually named after deceased, white, male officers and administrators) with terms that are more locally resonant and relevant.[7] In 2014, the City of Vancouver

---

5   Aaron Barcant, 'Language and Power!' *Convergence*, 4 (2013), 46–54, at 51, http://convergencejournal.ca/archives/484

6   Michif Languages Conference, *The Michif Languages Project: Committee Report* (Winnipeg: Manitoba Metis Federation, 1985).

7   See Rima Wilkes, Aaron Duong, Linc Kesler, and Howard Ramos, 'Canadian University Acknowledgement of Indigenous Lands, Treaties, and Peoples', *Canadian*

released the thoughtfully-compiled ninety-one-page *First Peoples: A Guide for Newcomers.*[8] Growing out of the Vancouver Dialogues Project (2010–13), an initiative to create more opportunities for understanding between Aboriginal and immigrant communities, the guide addressed the need for clear information in simple language about the First Peoples in Vancouver. Four years later, and as part of Vancouver's efforts toward reconciliation, city leadership consulted with members of the Musqueam, Squamish, and Tsleil-Waututh nations on whose traditional, ancestral and unceded territories the urban metropolis now sits in order to introduce a series of Indigenous place names for prominent landmarks. The plaza adjacent to the Queen Elizabeth Theatre is now šxʷƛ̓exən Xwtl'a7shn, a name linked to the plaza's past use as a gathering place for the Walk for Reconciliation. The Vancouver Art Gallery's north plaza has been named šxʷƛ̓ənəq Xwtl'e7énḵ Square, which refers to a place for a cultural gathering such as a wedding or funeral.

Universities and colleges are likewise engaged in these decolonial acts of toponymy. In 2016, the degree-granting Langara College was gifted the traditional Musqueam name snəw̓eyəł leləm̓ meaning 'house of teachings', with the term snəw̓eyəł referencing advice that is given to children to guide them into adulthood and build their character. This was the first time a British Columbian First Nation bestowed an Indigenous name on a public, post-secondary institution, and the Musqueam name is visible on all Langara College signage and communications. The University of British Columbia (UBC), where I teach, and the Musqueam First Nation entered into a high-level Memorandum of Affiliation in 2006 to further the sharing of knowledge and the advancement of Musqueam and Aboriginal youth and adults in post-secondary education. This Affiliation agreement further strengthened the long-standing partnership between Musqueam and UBC's First Nations and Endangered Languages Program which was initiated in 1997 as part of the university's commitment to community-based collaboration with First Nations peoples. The primary purpose

Review of Sociology/Revue canadienne de sociologie, 54.1 (2017), 89–120, https://doi.org/10.1111/cars.12140, for an illuminating discussion of the five general types of acknowledgement.

8   Kory Wilson and Jane Henderson, *First Peoples: A Guide for Newcomers* (Vancouver: City of Vancouver, 2014), https://vancouver.ca/files/cov/First-Peoples-A-Guide-for-Newcomers.pdf

of the partnership has been to promote the development and use of həṅ̓q̓əmiṅ̓əm̓, the Musqueam Central Coast Salish language, through collaborative research and teaching initiatives. In the nearly two decades since it began, the partnership has produced several formal research papers, a series of elementary resource books, and a full complement of text and interactive online materials that support four levels of həṅ̓q̓əmiṅ̓əm̓ language courses for post-secondary credit. These courses are open to both university and Musqueam students and serve as a powerful model for reconciliation.

My university has an uncomfortable history when it comes to relations with Indigenous communities and the ethics of appropriation. Totem Park Residence, a first-year dormitory that houses over 2,000 students, comprises eight houses with Indigenous 'names'. The original six house names — Nootka, Dene, Haida, Salish, Kwakiutl, and Shuswap — were selected in the 1960s without community consultation. As Sarah Ling compellingly argues, while they 'were intended to honor local BC First Nations', these names, 'along with intellectual property of the communities they represent, have long been misused, misrepresented, and appropriated due to a lack of context and education provided about them. Many of these names are also misnomers'.[9] In 2011, in a process led by Sarah Ling and Spencer Lindsay, both undergraduate students at the time, two new student residences were gifted names significant to the Musqueam Nation — həṁləsəṁ and q̓ələχən — through a collaborative and respectful process that incorporated Indigenous protocol and provided rich learning opportunities for student residents.

While such examples are exciting and inspiring, in order to make sense of contemporary efforts to revitalize Indigenous languages and cultural learnings, we need to understand the political and historical context that has shaped their marginalization. The use of the prefix 're' in words such as revitalization, rejuvenation, revival, and resurgence points to the undoing of some past action or deed.[10] If the world's linguistic diversity had not been 'devitalized' to begin with — through colonization, imperial adventure, war, and forced migration — there

---

9   Sarah Ling, 'həṁləsəṁ and q̓ələχən House Films Released!', 19 February 2014, https://ctlt.ubc.ca/2014/02/19/həṁləsəṁ-and-q̓ələχən-house-films-released/

10   Aaron Glass, 'Return to Sender: On the Politics of Cultural Property and the Proper Address of Art', *Journal of Material Culture*, 9.2 (2004), 115–39, https://doi.org/10.1177/1359183504044368

would be less need for historically marginalized languages with ever-dwindling numbers of speakers to be 'revitalized' today.

The work of language revitalization is inherently multidisciplinary and political, with long-range cultural and social goals that extend beyond the immediate task of generating more speakers. Increasingly, language revitalization programs are as much focused on decolonizing education and plotting a path toward Indigenous self-determination as they are directed at reclaiming grammar and speech forms. As Eve Tuck and K. Wayne Yang point out in their foundational contribution, 'Decolonization brings about the repatriation of Indigenous land and life; it is not a metaphor for other things we want to do to improve our societies and schools'.[11]

Language loss does not occur in isolation, nor is it inevitable or in any way 'natural'. The process also has wide-ranging social and economic repercussions for the language communities in question. Language is so heavily intertwined with cultural knowledge and political identity that speech forms serve as meaningful indicators of a community's vitality and social well-being. More than ever before, there are vigorous and collaborative efforts underway to reverse the trend of language loss and to reclaim and revitalize endangered languages. Such approaches vary significantly, from making use of digital technologies in order to engage individual and younger learners to community-oriented language nests and immersion programs. Drawing on diverse techniques and communities, the question of measuring the success of language revitalization programs has driven research forward in the areas of statistical assessments of linguistic diversity, endangerment, and vulnerability. Current efforts are re-evaluating the established triad of documentation-conservation-revitalization in favor of more unified, holistic, and community-led approaches.

The growing recognition of the legacy of colonial oppression of Indigenous languages has also motivated a realignment of the discourse around language endangerment. The majority of languages spoken across the world have endured punitive policies that actively sought to eradicate them. Their continued use to this day — even if only by a handful of speakers in some cases — is indicative of the resilience

---

11   Eve Tuck and K. Wayne Yang, 'Decolonization is Not a Metaphor', *Decolonization: Indigeneity, Education & Society*, 1.1 (2012), 1–40.

of communities in the face of continued oppression. Commonly used terms that highlight the 'endangered-ness' of a language — we may think of words such as 'weak', 'loss', and even the word 'endangered' itself — overrepresent diminishment and underrepresent the resurgent strength of communities of speakers who have never stopped using their ancestral languages.

The currency of terms such as 'vanishing' and 'disappearing' not only forecloses the possibility of revival and renewal but communicates an apparently agentless process in which language loss is both inevitable and naturally occurring. Such terminology both effaces the intentionality of colonial policies that legislated marginalization and undermines the efforts of those working to reclaim their languages. When speaking and writing of 'endangered languages', then, it is crucial to remain attentive to the words that are used and to seek balance in highlighting ongoing community revitalization efforts on the one hand, while historically contextualizing the increasingly vulnerable state of most Indigenous languages on the other.

With language reclamation and revitalization increasingly situated as an expression of self-determination and political empowerment, some language communities are developing a terminology for discussing endangerment and revitalization that is in itself empowering. One example is a movement to refer to languages without any current native or first-language speakers as 'sleeping' rather than 'extinct'.[12] While the distinction might appear unnecessary or even naively aspirational to researchers not closely involved in such work, all terminology has both symbolic value and political impact. The biological extinction of a species has a mono-directional finality that linguistic 'extinction' does not. As Indigenous linguist Wesley Leonard poignantly notes, 'the paradox of speaking an extinct language is not imaginary'.[13] The designation 'sleeping' rather than 'extinct' points to the potential of a language to be reclaimed and revived after it has lost its last

---

12 Leanne Hinton, 'Sleeping Languages: Can They Be Awakened?', in *The Green Book of Language Revitalization in Practice*, ed. by Leanne Hinton and Kenneth Hale (Leiden: Brill, 2001), pp. 411–17, https://doi.org/10.1163/9789004261723_032

13 Wesley Y. Leonard, 'When Is an "Extinct Language" Not Extinct?', in *Sustaining Linguistic Diversity: Endangered and Minority Languages and Language Varieties*, ed. by Kendall A. King et al. (Washington, DC: Georgetown University Press, 2008), pp. 23–33, at p. 28.

first-language speakers — an opportunity that is not available to the dodo or a dinosaur. While bringing a language back from sleeping to having a community of fluent speakers is a phenomenon that has been uncommon in human history, there are recent examples, such as the remarkable and compelling case of the Wampanoag (Algonquin) language, which was sleeping from the late nineteenth century until revitalization efforts resulted in fluent child speakers of the language in the twenty-first century.[14]

For peoples like the Myaamia (Algonquin), who have no first language speakers left, 'the ultimate goal of this work is to eventually be able to raise our children with the beliefs and values that draw from our traditional foundation and to utilize our language as a means of preserving and expressing these elements'.[15] Rather than some ideal, end-state fluency, it is the sustained effort of communities that shape and determine the goal and success of any language revitalization project. As all who are engaged in language revitalization can attest, the work is never complete: success starts when revitalization efforts begin and doesn't end until efforts themselves cease.[16]

Elders and youth in Indigenous communities are actively using and harnessing emerging technologies to strengthen their traditions and languages; Indigenous peoples are creators and innovators (not just recipients or clients) of new technologies, particularly in the domain of cultural and linguistic heritage. While technological efforts in the 1970s included specially modified typewriters and custom-made fonts to represent Indigenous writing systems, communities are now making use of digital tools — online, text, Internet radio and mobile devices — to nurture the continued development of their respective diverse Indigenous languages and cultures. Yet, such interventions are not without risks and consequences. Digital technologies cannot and will

---

14   *We Still Live Here: Âs Nutayuneân*, dir. by Anne Makepeace (2010).

15   Daryl Baldwin, 'Miami Language Reclamation: From Ground Zero', *Lecture Presented at the 24th Speaker Series at the Center for Writing* (Minneapolis, MN: University of Minnesota, 2003), p. 28, http://writing.umn.edu/lrs/assets/pdf/speakerpubs/baldwin.pdf

16   Leanne Hinton, 'Leanne Hinton: What Counts as "Success" in Language Revitalization?', 55:44, posted online by The University of British Columbia, *Youtube*, 3 November 2015, https://www.youtube.com/watch?v=qNlUJxri3QY. This talk was part of the Future Speakers: Indigenous Languages in the 21st Century Series, held at the University of British Columbia, Vancouver, 2015.

not save languages. Speakers keep languages alive. A digital dictionary itself won't revitalize an endangered language, but could assist the speakers who will. At the same time, technology can be as symbolically powerful as it is practically useful, and can carry considerable political weight. In the English-dominant world of cyberspace, Indigenous communities are engaging with, disrupting, and re-imagining digital practices. By generating digital visibility and legibility, Indigenous communities claim a 'presence' online, and exert control over the terms of Indigenous representation rather than risk misrepresentation.

As a practice, language revitalization takes many shapes. Some of the earliest language activists were the children and students who, risking corporal or psychological punishment, continued to speak their languages in residential and boarding schools and at home with their families. Since the retraction of explicit bans on speaking Indigenous languages in public in many countries, some of which have only been lifted within the last few decades, language revitalization has become noticeably less subversive.[17] Many language revitalization programs now receive support from band councils, non-governmental organizations, philanthropic foundations, and even governmental bodies and programs.

Recalling the central relevance of language to many other aspects of community well-being, the transformative healing nature and holistic benefits of language revitalization have much wider impact and relevance than linguistic vitality alone.[18] Indigenous language revitalization speaks as much to 'hard' indicators of health and well-being as it does to 'soft' indicators of culture and identity. As the Sto:lo/ Xaxli'p educator and writer Q'um Q'um Xiiem (Jo-ann Archibald) said to Aboriginal educators at *Oral Traditions: The Fifth Provincial Conference on Aboriginal Education* in 1999, while 'we need to preserve

17 Eric A. Anchimbe, 'Functional Seclusion and the Future of Indigenous Languages in Africa: The Case of Cameroon', in *Selected Proceedings of the 35th Annual Conference on African Linguistics*, ed. by John Mugane et al. (Somerville, MA: Cascadilla Proceedings Project, 2006), pp. 94–103; Mekuria Bulcha, 'The Politics of Linguistic Homogenization in Ethiopia and the Conflict Over the Status of "Afaan Oromoo"', *African Affairs*, 96.384 (1997), 325–52, https://doi.org/10.1093/oxfordjournals.afraf.a007852

18 D. H. Whalen, Margaret Moss, and Daryl Baldwin, 'Healing through Language: Positive Physical Health Effects of Indigenous Language Use', *F1000Research*, 5 (2016), 852, https://doi.org/10.12688/f1000research.8656.1

our oral traditions, we also need to let them preserve us'. Important new studies demonstrate the interrelatedness of language and community well-being. A recent Canadian study showed a compelling correlation between Indigenous language use and a decrease in Aboriginal youth suicide rates in British Columbia.[19] Such statistical research helps to highlight the multidimensional nature of language revitalization and its cross-sector impact on the lives and livelihoods of Indigenous communities.

At the same time, we need to situate language work in the wider context of biocultural diversity, which Luisa Maffi helpfully defines as 'the diversity of life in all its manifestations: biological, cultural, and linguistic, which are interrelated (and possibly coevolved) within a complex socio-ecological adaptive system'.[20] Over the last decades, researchers in previously unrelated fields have begun to explore exciting correlations between human and biological worlds, specifically in relation to language. There is an emerging consensus between scientists and humanists that biodiversity and linguistic diversity go hand-in-hand: areas rich in one are usually rich in the other. Scholarship in this field emphasizes that the diversity of life comprises not only the variety of species and cultures that have evolved on earth, but also the diverse human languages that have developed over time. An integrated biocultural approach highlights the importance of languages in not only the communication and transmission of cultural values, but also in maintaining traditional knowledge and ecological practices. By extension, a biolinguistic perspective argues for the centrality of language in mediating human-environment interactions and mutual adaptations.

The Federal Government of Canada and its research councils are beginning to provide targeted resources to explore the intersection of language, well-being and health. Prime Minister Justin Trudeau spoke to the Assembly of First Nations in December 2016, pledging to introduce a federal law to protect, preserve, and revitalize First Nations, Inuit,

19   Hallet, Darcy, Michael J. Chandler, and Christopher. E. Lalonde, 'Aboriginal Language Knowledge and Youth Suicide', *Cognitive Development*, 22 (2007), 392–99, https://doi.org/10.1016/j.cogdev.2007.02.001

20   Luisa Maffi, 'Biocultural Diversity and Sustainability', in *The SAGE Handbook of Environment and Society*, ed. by Jules Pretty et al. (Thousand Oaks, CA: SAGE, 2007), pp. 267–77, at p. 269, https://doi.org/10.4135/9781848607873.n18

and Métis languages: 'We know [...] how residential schools and other decisions by government were used to eliminate Indigenous languages. We must undo the lasting damage that resulted [...] Today I commit to you our government will enact an Indigenous Languages Act'. Working with leaders from First Nations communities who have been advocating and calling for such legislation for decades, the Trudeau Government finally introduced Bill C-91, An Act Respecting Indigenous Languages, into law in 2019.

The bitter irony of the current context is inescapable: colonial governments have since colonization marshaled their economic, military, and administrative might to extinguish Indigenous voices. Now, in the eleventh hour, they are looking to resource that which they first set out to destroy. Benign neglect would have been less damaging than two centuries of violence followed by a last-minute U-turn. Will citizens of settler colonial nations hold their government to account and demand that effective and progressive Indigenous languages legislation be enacted?

We need to listen to and learn from Indigenous communities, honor their processes and goals, and support community-led revitalization programs through respectful partnership. Indigenous communities know their needs better than anyone, and acknowledging this place-based expertise is a step towards reconciliation. Indigenous communities need better resourcing for language instructors to promote stronger learning outcomes, language retention and trust. Indigenous communities must be supported to set their own goals, as these are more attainable, more credible and have a higher chance of fulfillment than any imposed from outside. Indigenous communities need more funding, dispersed in a better way, to plan strategically over the long term. Communities must not be positioned as competitors for resources and visibility, but rather have dedicated funding streams that will enable long-term sustainability.

As ever, leadership is coming from the grassroots. From September 2015, all students in kindergarten through Grade 4 (ages 10–11) in Prince Rupert, British Columbia, have been learning Sm'algyax. 'We are on traditional Tsimshian territory and Sm'algyax is the language of the territory', Roberta Edzerza (Aboriginal Education Principal for her District) told CBC Radio One. 'We are so proud and we would like to share our language and culture with everybody. It's one avenue to address racism. Education is key. Learning the language and sharing in the learning and the culture'.[21]

While the alarms bells have sounded and the threat of languages ceasing to be spoken remains a reality for increasing numbers of communities, the indomitable human spirit in the face of adversity should not be underestimated. Language communities across the globe have proven throughout history that the odds *can* be beaten and that the effects of colonization are surmountable. Indigenous communities need dedicated and longterm resources to design and implement their own research agendas, learning goals, and success criteria for language revitalization and reclamation work. Through engaging in collaborative linguistic and cultural revitalization work, building partnerships, and nurturing communities of practice at academic, governmental, and grassroots levels, the tide can be turned and more languages will join the ranks of Hawaiian, Māori, Myaamia, and Wampanoag.

Bibliography

Anchimbe, Eric A., 'Functional Seclusion and the Future of Indigenous Languages in Africa: The Case of Cameroon', in *Selected Proceedings of the 35th Annual Conference on African Linguistics*, ed. by John Mugane et al. (Somerville, MA: Cascadilla Proceedings Project, 2006), pp. 94–103.

Baldwin, Daryl, 'Miami Language Reclamation: From Ground Zero', *Lecture Presented at the 24th Speaker Series at the Center for Writing* (Minneapolis, MN: University of Minnesota, 2003), http://writing.umn.edu/lrs/assets/pdf/speakerpubs/baldwin.pdf

Barcant, Aaron, 'Language and Power!', *Convergence*, 4 (2013), 46–54, http://convergencejournal.ca/archives/484

Bulcha, Mekuria, 'The Politics of Linguistic Homogenization in Ethiopia and the Conflict Over the Status of "Afaan Oromoo"', *African Affairs*, 96.384 (1997), 325–52, https://doi.org/10.1093/oxfordjournals.afraf.a007852

Glass, Aaron, 'Return to Sender: On the Politics of Cultural Property and the Proper Address of Art', *Journal of Material Culture*, 9.2 (2004), 115–39, https://doi.org/10.1177/1359183504044368

---

21   'Students in Prince Rupert to Learn Indigenous Language', *Daybreak North*, CBC Radio One, 9 June 2015, https://www.cbc.ca/news/students-in-prince-rupert-to-learn-indigenous-language-1.3108265

Hallet, Darcy, Michael J. Chandler, and Christopher. E. Lalonde, 'Aboriginal Language Knowledge and Youth Suicide', *Cognitive Development*, 22 (2007), 392–99, https://doi.org/10.1016/j.cogdev.2007.02.001

Hinton, Leanne, 'Sleeping Languages: Can They Be Awakened?', in *The Green Book of Language Revitalization in Practice*, ed. by Leanne Hinton and Kenneth Hale, pp. 411–17 (Leiden, The Netherlands: Brill, 2001), https://doi.org/10.1163/9789004261723_032

— 'Leanne Hinton: What Counts as "Success" in Language Revitalization?', 55:44, posted online by The University of British Columbia, *Youtube*, 3 November 2015, https://www.youtube.com/watch?v=qNlUJxri3QY.

Krauss, Michael, 'The World's Languages in Crisis', *Language*, 68.1 (1992), 4–10, https://doi.org/10.1353/lan.1992.0075

Leonard, Wesley Y., 'When Is an "Extinct Language" Not Extinct?', in *Sustaining Linguistic Diversity: Endangered and Minority Languages and Language Varieties*, ed. by Kendall A. King et al. (Washington, DC: Georgetown University Press, 2008), pp. 23–33.

Ling, Sarah, 'hǝṁlǝsǝṁ and q̓ǝlǝχǝn House Films Released!', 19 February 2014, https://ctlt.ubc.ca/2014/02/19/hǝṁlǝsǝṁ-and-q̓ǝlǝχǝn-house-films-released/

Maffi, Luisa, 'Biocultural Diversity and Sustainability', in *The SAGE Handbook of Environment and Society*, ed. by Jules Pretty et al. (Thousand Oaks, CA: SAGE, 2007), pp. 267–77, https://doi.org/10.4135/9781848607873.n18

Michif Languages Conference, *The Michif Languages Project: Committee Report* (Winnipeg: Manitoba Metis Federation, 1985).

Pine, Aidan, and Mark Turin, 'Language Revitalization', in *Oxford Research Encyclopedia of Linguistics*, ed. by Mark Aronoff (New York, NY: Oxford University Press, 2017), https://doi.org/10.1093/acrefore/9780199384655.013.8

'Students in Prince Rupert to Learn Indigenous Language', *Daybreak North*, CBC Radio One, 9 June 2015, https://www.cbc.ca/news/students-in-prince-rupert-to-learn-indigenous-language-1.3108265

Sutherland, William James, 'Parallel Extinction Risk and Global Distribution of Languages and Species', *Nature*, 423.6937 (2003), 276–79, https://doi.org/10.1038/nature01607

Tuck, Eve, and K. Wayne Yang, 'Decolonization is Not a Metaphor', *Decolonization: Indigeneity, Education & Society*, 1.1 (2012), 1–40.

We Still Live Here: Âs Nutayuneân, dir. by Anne Makepeace (2010).

Whalen, D. H., Margaret Moss, and Daryl Baldwin, 'Healing through Language: Positive Physical Health Effects of Indigenous Language Use', *F1000Research*, 5 (2016), 852, https://doi.org/10.12688/f1000research.8656.1

Wilkes, Rima, Aaron Duong, Linc Kesler, and Howard Ramos, 'Canadian University Acknowledgement of Indigenous Lands, Treaties, and Peoples', *Canadian Review of Sociology/Revue canadienne de sociologie*, 54.1 (2017), 89–120, https://doi.org/10.1111/cars.12140

Wilson, Kory, and Jane Henderson, *First Peoples: A Guide for Newcomers* (Vancouver: City of Vancouver, 2014), https://vancouver.ca/files/cov/First-Peoples-A-Guide-for-Newcomers.pdf

Wurm, S., Muehlheausler, P., and D. Laycock, 'Language Planning and Engineering in Papua New Guinea', *New Guinea Area Language and Language Study*, 3 (1977), 1157–77.

16. Sensing, Minding, and Creating

John Grim

Sensing, minding, and creating are integrated ways that the human perceives, thinks about, and finds novel pathways in the immensity of the world, and, more immediately, with the 'tangled bank' of life. I have used two different styles to notate this: namely, *sensing-minding-creating* (connected by dashes), and *sensing, minding, and creating* (distinguished by commas). Each style refers to this process of universe emergence (please note that here I am aligning this happening in the human with the happening of the world). *Sensing-minding-creating* emphasizes, through connective dashes, the simultaneity and interwoven character of this threefold interaction; whereas the use of commas emphasizes the distinctive manifestations of the three processes that make up this interaction. Although the integrated actions of all three are simultaneous and interactive dimensions, sensing is placed first, and thus emphasized, to draw attention to the world's multisensory experience of itself as foundational to existence. That is, sensing, or 'reaching out', characterizes both inorganic and organic existence, and minding, or 'inner patterning' or 'consciousness', characterizes all reality from the primal flaring forth of our universe. Creating, or emerging, follows from the bending back of sensing-minding on themselves giving rise to novel space for existence to flourish, change, and evolve.

Sensing and the woven fabric of sensations occurs, for example, in the need for material forms to bend back towards and into one another. In the lifeworld (the whole fabric of reality expressed in individual subjective responses to the world), sensing cells discern, through their

 https://doi.org/10.11647/OBP.0186.16

reaching out, that which also enables pattern and flourishing. Patterning or consciousness (what I call minding) arises in the formation of galaxies, in the need for cells to discern, and in human reaching out in thought. Creating results from that reaching out and discerning, which returns, or bends back, in ways that open new possibilities for patterning and flourishing.

In some cultural modes, the role of sensing has been considered as ancillary to minding and creating. In this exploratory statement (sensing-minding-creating), I seek to reposition our sensing of the world as central to understanding the human. Moreover, I want to emphasize ways in which humans, following material forms in the emergence of the universe, and the evolution of life, have been brought to new places of questioning and knowing by the world.

My own particular concerns for the study of religion and ecology opens an awareness in me of ways in which the human becomes attentive to the world as both integral to itself, and other-than itself, through sensing-minding-creating. I focus here on religion and ecology as an approach for understanding sensing-minding-creating, not as a final or ultimate exemplar of these integrated dimensions. Rather, the study of religion and ecology explores human experience as a bridging of inner experiences and external relatedness to a world perceived as filled with sacred meaning. Religions in various ways acknowledge the integral dependence of all of life's communities on a larger whole. Now, through the empirical search of modern science, we learn of life's dependence on Earth's ecosystems. This is an historical dependence of reaching out in need and patterning our responses. This is a dependence that arises from what came before, but which gives rise to radically new emergent forms among non-living and living existence.

This approach to studying religion and ecology explores the ways in which humans interrogate the world. This is a search, as I see it, for ultimate transcendence within the boundaries of existence in the universe. Human encounters with the natural world, as well as engagements with their own social and built worlds, all stand in relation to one another as sensed realities that give rise to thought, doubt, and question. Perception of the world via our senses arouses commitments that the world is as our sensations reveal. Yet, that commitment to perception continually flounders as our senses themselves rise up against our efforts to order rationally the changing world.

The paleolithic cave paintings, grave offerings and extensive feasting of our hominin ancestors demonstrate a reaching out into the world some 30,000 years before the present (BP), to 2 million years BP. Wouldn't we also say that that life manifest in the ordered styles of paleolithic cave paintings is also filled with doubt, contestation, and question? All this questioning has not simply occurred in some speculative realm, but in direct relation to our ancestors' lived-experience of the celestial realms, seasons, landscapes, and in the migratory patterning of biodiversity. These are the relational manifestations of sensing-minding-creating.

This questioning begins with human space — that is, the space of the human person in the world, a space in which the search for religious ultimate presence and meaningful pathways appear. This human space cannot be separated from the surrounding world. Transformed by the mutual perceiving of self, world, and other, a neutral space — formerly empty and formless — emerges. The term hierophany has been used in religious studies to suggest the human experience of the sacred, which is a reflexive place of sensing-minding-creating.

Religious hierophanies are examples, then, of the ways that somatic sensing reaches out, somatic thinking orders in patterns of meaning, and, emerging in the world, opens into an emptiness capable of flourishing. In these ways the human and the world come to understand and clarify themselves. In this process, they continue to reveal and to question the deeper structures of reality in relation to each other. Truths are composite narratives expressed in the sensing-minding of existent beings subject to constantly change. This change, arising from out of the inherent foaming into existence of the particulate world, is charged by that foaming emptiness. In this charge a dynamic bending occurs in sensing-minding back upon themselves into the foaming emptiness that is also towards creating.

For the human, sensing-minding-creating are responses within the world as a communion of subjects.[1] In the study of religion and ecology, sensing-minding-creating presents a 'process approach' (in which the wholeness of universe processes predominates); whereas sensing, minding, and creating present an attempt to preserve the need

---

1   The idea of the world as a 'communion of subjects', as opposed to a 'collection of objects', is at the heart of Thomas Berry's work: see, for example, *The Dream of the Earth* (San Francisco and Berkeley, CA: Sierra Club Books, 1988).

among existent beings for individual subjectivity. In these mutually aligned frames (sensing-minding-creating), the diverse religions can be interrogated regarding: (i) the sense of the human and the world for each other, which leads to an awareness of their mutual kinship and their dependence on each other; (ii) the capacity of human intelligence to frame, conceptually, the sensible world so as to creatively renew itself in ultimate transcendence; (iii) the human quest to return creatively to the revelatory place of ultimate transcendence for transformation even as the flow of perceived events carries their world forward.

Sensing, minding, and creating are presented as specific modes of questioning for understanding the diverse religions in their formative and ongoing contacts with different ecosystems. Religions are, after all, composed and transmitted by human beings with bodies who experience the world, and who, according to the particularity of religious traditions, experience a revelatory message with regard to the world. That aspect of religion, namely, its relevance to the world, is a primary consideration here, though it may not be situated as primary within a given tradition. For example, Torah was given to the Jewish community in both oral and literate forms as a questioning-knowing covenant between the Divine, the Creation, and the Chosen people. Among Indigenous Tewa peoples, for example, of San Juan Pueblo, summer lineages interact and vie with winter lineages in questioning-knowing relationships. Native Koyukon peoples speak of a watchful world that may obscure meaning in perception, yet reveals in an oral tradition of ethical stories of the 'long ago'. Confucian-Daoist sages suggest interactions with the Dao may reveal interior landscapes with cosmological implications. The ground upon which these questioning-knowing relationships occur in the world is a way of interpreting these covenant-community relationships.

Sensing, minding, and creating provide us with interrogative approaches appropriate even for the seemingly transcendent orientations of religious traditions. A transcendence, I sense, whether it be heaven-oriented or not, is still reaching out from bodies for novel flourishing. This approach, then, acknowledges the questioning process at the heart of our human experience of a changing world, even as it creatively searches for that abiding place where ultimate knowing resides. In this sense, our approach is not a quest for philosophical insight as much as an opening to the multiple ways in which ecosystems and world-pictures have affected the quests of religious life. Acknowledging the seminal roles of empirical observation, or sensing, and of thought, or minding, in the emergence of religions, this approach also emphasizes the particular turns in religions we call creating. One move that appears evident in the religions is the bending back of sensing and minding that effects a doubling in the flow that opens out into creative mystery.

The intention of this approach is to bridge the divide that has been established between the human and the natural world, culture and nature. Sensing the world, and attention to the world as sensing the human, are presented as a unifying ground, that place from which minding, or thought about the world, meets itself. Such an opening to religions as complex interactions of sensing-minding-creating presumes experiencing bodies, thinking minds, and creative engagement on both sides of the encounter. Such a generative place is not without doubt and questioning. But it is also a shared ground, and a sense of shared bodies gives perceptual depth to doubt and question. It is the case that anthropological studies have described cultures in which distinctions are made between the realm of human society and the nonhuman world.[2] While not ignoring these meaningful, symbolic, pragmatic distinctions, a more radical nature/culture separation may be a modern turn. Thus, a deeper, human sense of difference from ecology may pervade our religions, and this merits reconsideration. As the Indigenous scholar Jack Forbes writes, 'People can be, and indeed are, part of "nature." The objective is to understand that together-doing of the balanced kind and the Away from People have never been mutually exclusive and that "nature needs people" (just not too many of them!)'.[3]

This differs from a variation on an older, comparative religions approach that is often identified with Friedrich Max Müller. In that

---

2   My analysis here is informed by Claude Lévi-Strauss and his structural attention to dualisms of culture and nature; Alphonso Ortiz's studies of the dualisms within Tewa/San Juan Pueblo culture (*The Tewa World: Space, Time, Being, and Becoming in a Pueblo Society* (Chicago, IL: University of Chicago Press, 1969)); Jack Forbes, 'Nature and Culture: Problematic Concepts for Native Americans', in *Indigenous Traditions and Ecology: The Interbeing of Cosmology and Community* (Cambridge, MA: Harvard University Press for the Center for the Study of World Religions, 2001), pp. 103–24; and the distinctions made by Richard Nelson throughout *Make Prayers to the Raven: A Koyukon View of the Northern Forest* (Chicago, IL: University of Chicago Press, 1986).

3   Forbes, 'Nature and Culture', p. 122.

nineteenth-century view, religious thought and language relate to ecological realities as corrupted remnants of lived-experiences in the world. From this perspective, thought, or minding, would faintly echo the sensed, experienced world. The position taken in this chapter, and presented in the process approach of sensing-minding-creating, emphasizes the seminal role of creative doubt and questioning in the lived experience of the world. Rather than experiences of the natural world frozen in language, religions transmit the capacity for ongoing experiences and expressions of body-mind in the intertwined worlds of culture-nature. Yet, the established patterning guiding a study of mutual influences of bioregions and religions is often masked in the mystery of bending back upon themselves. Certainly, patterns are developed in the traditions themselves for expressing places of ultimacy and transformation. It also evident that historical studies describe the many ways in which religions relate to ecosystems. These can be described using categories from scientific ecology and even more pointedly using terms drawn from the particular religions themselves. What this chapter seeks to elucidate here is that the underlying dynamics that have generated such patterns of interaction between religions and ecologies mask as much as they reveal. Their contemporary relevance, and the ongoing doubt and questioning they present, are subject to our own rising up into that emptiness holding the potentiality for flourishing.

While Bertrand Russell, in his autobiography, stood at the edge of a dark ocean crying into the night of his fading nineteenth-century Protestantism, the larger prospect of all the human religious communities may be more poignantly described by the chasm we have created between selves and the world.[4] Climate emergency, and the larger environmental diminishment, confront us like a chasm between ourselves and all that we have known. As W. G. Sebald described in his novel *Austerlitz*, it is 'truly terrifying to see such emptiness open up a foot away from firm ground, to realize that there was no transition, only this dividing line, with ordinary life on one side and its unimaginable opposite on the other. The chasm into which no ray of light could penetrate...'[5]

---

4   See the years 1951–69 in Bertrand Russell, *The Autobiography of Bertrand Russell, Vol. 2* (London: George Allen & Unwin, 1956).

5   W. G. Sebald, *Austerlitz*, trans. by Anthea Bell (New York, NY: Modern Library/ Random House, 2001), p. 297.

Bibliography

Berry, Thomas, *The Dream of the Earth* (San Francisco and Berkeley, CA: Sierra Club Books, 1988).

Forbes, Jack, 'Nature and Culture: Problematic Concepts for Native Americans', in *Indigenous Traditions and Ecology: The Interbeing of Cosmology and Community*, ed. by John Grim (Cambridge, MA: Harvard University Press for the Center for the Study of World Religions, 2001), pp. 103–24.

Nelson, Richard, *Make Prayers to the Raven: A Koyukon View of the Northern Forest* (Chicago, IL: University of Chicago Press, 1986).

Ortiz, Alphonso, *The Tewa World: Space, Time, Being, and Becoming in a Pueblo Society* (Chicago, IL: University of Chicago Press, 1969).

Russell, Bertrand, *The Autobiography of Bertrand Russell, Vol. 2* (London: George Allen & Unwin, 1956).

Sebald, W. G., *Austerlitz*, trans. by Anthea Bell (New York, NY: Modern Library/ Random House, 2001).

17. Land, Indigeneity, and Hybrid Ontologies

Paul Berne Burow, Samara Brock, and Michael R. Dove[1]

Introduction

How we understand land is fundamentally about how we see ourselves and our relationship to human and non-human others. In this piece, which is an excerpt from a larger essay, perspectives from political ecology, posthumanism, and Indigenous studies are combined to illuminate the importance of understanding land as something that transcends the material and is deeply enmeshed with our self-identity and relationality. It concludes that seeing that possible multiple, or hybrid, ontological framings of land can co-exist can have political power especially in relation to Indigenous movements for social and environmental justice.

Political Ecology, Land, and Environmental Subjectivities

Issues of land have always been central to political ecology scholarship. Early work in political ecology examined the issue of land degradation, calling attention to the political-economic forces that work to discursively produce land according to socially constructed schema of quality, rather

---

1   This is an excerpt from a longer paper, 'Unsettling the Land: Indigeneity, Ontology, and Hybridity in Settler Colonialism', by Paul Berne Burow, Samara Brock, and Michael R. Dove, *Environment and Society: Advances in Research*, 9 (2018), 57–74, https://doi.org/10.3167/ares.2018.090105

    https://doi.org/10.11647/OBP.0186.17

than reflecting an 'objective' ecological condition outside of society.[2] In recent years, this literature has challenged scholars to show the inextricable connection between how we see 'nature', and how power moves in relation to its management.[3] Tim Ingold outlines two opposing ways of seeing land as, on one hand, a spherical, embedded, localized perception and, on the other, a global view where the human transcends nature and the world is seen as property or resources to be managed for the public good.[4] This global, aerial view engenders a perspective that writers such as James Scott and Bruno Latour have similarly seen as enabling abstraction, measurement, calculation, and accumulation of knowledge by experts at centers of calculation and power.[5] These authors focus on the role of simplification in enabling power to expand, including the simplification of landscapes. These simplifications make things legible (as Scott argues) or immutable, accumulable, and combinable (as Latour argues). When the simplifications ultimately fail, this failure is not acknowledged as such, but rather understood as something in need of a techno-scientific fix.[6] Techno-scientific claims of understanding — and thus power over — landscapes are thus strengthened.

Through this process, the way in which landscapes are seen shifts. As Donald S. Moore observes in his analysis of struggles for territory in Kaerezi, Zimbabwe, 'abstract, empty, and exchangeable space is a historical product, not an essence'.[7] In his examination of the impacts of colonization in Egypt, Timothy Mitchell describes how the process facilitated 'the spread of a political order that inscribes in the social world a new conception of space, new forms of personhood, and a new

---

2   Piers Blaikie, *The Political Economy of Soil Erosion in Developing Countries* (London: Longman, 1985); Piers Blaikie and Harold Brookfield, *Land Degradation and Society* (London: Methuen, 1987).

3   Cf. Paul Robbins, *Political Ecology: A Critical Introduction* (New York, NY: John Wiley & Sons, 2012).

4   Tim Ingold, *The Perception of the Environment: Essays on Livelihood, Dwelling, and Skill* (London: Routledge, 2000), https://doi.org/10.4324/9780203466025

5   James C. Scott, *Seeing Like a State: How Certain Schemes to Improve the Human Condition Have Failed* (New Haven, CT: Yale University Press, 1998); Bruno Latour, *Science in Action: How to Follow Scientists and Engineers through Society* (Cambridge, MA: Harvard University Press, 1987).

6   Timothy Mitchell, *Rule of Experts: Egypt, Techno-Politics, Modernity* (Berkeley, CA: University of California Press, 2002).

7   Donald S. Moore, *Suffering for Territory: Race, Place, and Power in Zimbabwe* (Durham, NC: Duke University Press, 2005), p. 20, https://doi.org/10.1215/9780822387329

means of manufacturing the experience of the real'.[8] Thus, another key insight from the political ecology literature is that, whereas ways of understanding land can change how power over land operates, these concepts can also change actors' subjectivities, changing how they manage their own conduct in relation to land. Looking at colonialism, development, and other projects of land management, a number of authors, often influenced by Michel Foucault's work on governmentality, explore how subjects come to participate in projects of their own rule.[9] In Moore's words, 'different political technologies produce territory, including its presumed "natural, features"'.[10] Jeremy M. Campbell explores how settlers on the frontier can work to conjure private property in the absence of a strong state presence, demonstrating the limits of state-centric approaches that fail to account for the political and economic power of settlers to realize their own visions of a transformed landscape.[11] Bruce Braun examines how the evolution of a geological vision impacted conceptualizations of Haida Gwaii (formerly known as the Queen Charlotte Islands) in Canada.[12] Braun argues that when governing is done to manage the relationship between populations and territory, the qualities of territory (land) are not static, but rather are continuously reconstituted as a result. Governing must be continuously reordered to structure conduct in response to shifting constructions of nature. The adoption of a geological understanding of land in Haida Gwaii, for example gave rise to new forms of calculation and governance in relation to it. As land came to be understood as vertical, human subjectivities changed to better manage it. Arun Agrawal touches upon similar ideas in his discussion of community forestry programs in Kumaon, India. He conceives of environmentality as a framework

---

8   Timothy Mitchell, *Colonising Egypt* (Berkeley, CA: University of California Press, 1991), p. ix.

9   Arun Agrawal, *Environmentality: Technologies of Government and the Making of Subjects* (Durham, NC: Duke University Press, 2005), https://doi.org/10.1215/9780822386421; Tania Murray Li, *Land's End: Capitalist Relations on an Indigenous Frontier* (Durham, NC: Duke University Press, 2014), https://doi.org/10.1215/9780822376460

10   Moore, *Suffering for Territory*, p. 7.

11   Jeremy M. Campbell, *Conjuring Property: Speculation and Environmental Futures in the Brazilian Amazon* (Seattle, WA: University of Washington Press, 2015).

12   Bruce Braun, 'Producing Vertical Territory: Geology and Governmentality in Late Victorian Canada', *Cultural Geographies*, 7.1 (2000), 7–46, https://doi.org/10.1177/096746080000700102

for understanding how environmental subjects are created, through participation in the 'intimate government' of local forests.[13] Timothy Luke similarly shows how subjectivities of expert management are recast when nature is conceived in terms of coupled socioecological systems.[14] Seeing nature as a complex system under threat invites expert managerial control. By examining the work of three technical scientific bodies, Luke demonstrate how Earth System science has given rise to a global green governmentality exercised by international technical experts who map, monitor, measure, and, ultimately, manage nature and population for the public good.

For these authors, self-interest comes to be realized through participation in different forms of practice. For Tania Murray Li, ways of understanding land outline what, and especially who, is excluded from that land.[15] Every regime of exclusion has to be legitimated and can, therefore, be contested. Li notes the prominence of moral arguments and references to the social value of investment in driving contemporary land grabs. This is the extension of Ingold's idea of the need to optimize land use for the public good: not only can we manage land according to global understanding, but we must do so for the public good, even if some publics' interests must be sacrificed to do so.

This literature, though useful in examining how we come to understand land, has been critiqued for some of its limitations. Thus, the emergence of an Indigenous political ecology has built on these insights but also sought to address the elisions of political-economic approaches that are 'limited by a reliance on Euro-derived concepts of power, political economy and human — environmental relations [... that] may reproduce colonial relations of power, while eliding Indigenous peoples' own solutions to problems'.[16] The unique position

---

13   Agrawal, *Environmentality*, p. 178.

14   Timothy W. Luke, 'Developing Planetarian Accountancy: Fabricating Nature as Stock, Service, and System for Green Governmentality', in *Nature, Knowledge, and Negation*, ed. by Harry Dahms (Bingley, UK: Emerald Group Publishing Limited, 2009), pp. 129–59, https://doi.org/10.1108/S0278-1204(2009)0000026008

15   Tania Murray Li, 'What is Land? Assembling a Resource for Global Investment', *Transactions of the Institute of British Geographers*, 39.4 (2014), 589–602, https://doi.org/10.1111/tran.12065

16   Beth Rose Middleton, 'Jahát Jatítotòdom: Toward an Indigenous Political Ecology', in *The International Handbook of Political Ecology*, ed. by Raymond Bryant (Northampton, MA: Edward Elgar Publishing, 2015), pp. 561–76, at p. 561, https://doi.org/10.4337/9780857936172.00051

of Indigenous peoples, given their status as both authorities on their homelands at the same time they are subjects of a settler state which lays claim to this homeland, contrasts with many cases in political ecology where 'singular states comprise the operational governmental authority to which their subjects must react'.[17] Beth Rose Middleton outlines the key tenets of an Indigenous political ecology as:

> (1) attention to 'coloniality' or ongoing practices of colonialism (e.g. displacement of indigenous peoples from their lands; no recognition of indigenous self-determination); (2) culturally specific approaches reframing analyses in keeping with indigenous knowledge systems; (3) recognition and prioritization of indigenous self-determination, as expressed through indigenous governance; and (4) attention to decolonizing processes that explicitly dismantle systems of internalized and externalized colonial praxis.[18]

Conceptions of land configure how one relates, not just to land, but to many other actors — human and nonhuman — in the broader community.[19] In accepting colonial recognition of their rights to land, Indigenous nations can end up undermining their reciprocal relationships to that land. Glen Coulthard argues that political recognition of Indigenous peoples in Canada obscures the ongoing settler colonial project of primary accumulation — the drive toward dispossession of Indigenous lands while extracting further surplus value through resource exploitation.[20] And that any attempt to transcend these structures of domination requires the resuscitation of relationships of mutual obligation between land and people as opposed to deeper engagement with settler-state institutions. Clint Carroll, writing about the Cherokee Nation, also remarks that Indigenous environmental governance represents a different, 'relationships-

---

17   Clint Carroll, 'Native Enclosures: Tribal National Parks and the Progressive Politics of Environmental Stewardship in Indian Country', *Geoforum*, 53 (2014), 31–40, at 37, https://doi.org/10.1016/j.geoforum.2014.02.003

18   Middleton, 'Jahát Jatítotòdom', p. 562.

19   Paul Nadasdy, *Hunters and Bureaucrats: Power, Knowledge, and Aboriginal-State Relations in the Southwest Yukon* (Vancouver: UBC Press, 2003); Paul Nadasdy, 'The Gift in the Animal: The Ontology of Hunting and Human-Animal Sociality', *American Ethnologist*, 34.1 (2007), 25–43, https://doi.org/10.1525/ae.2007.34.1.25

20   Glen Coulthard, *Red Skin, White Masks: Rejecting the Colonial Politics of Recognition* (Minneapolis, MN: University of Minnesota Press, 2014), https://doi.org/10.5749/minnesota/9780816679645.001.0001

based approach' that allows for 'agency of nonhuman beings and the maintenance of relationships with them'.[21]

Recent critiques of the cultural underpinnings of sovereignty inherent to Indigenous nation-building have suggested that sovereignty itself can have problematic instrumental effects.[22] But the necessity of engaging its forms still stands. One hybrid approach is the use of land-as-property in creative ways that inflect its forms to promote the creation of Indigenous space. As Carroll notes, 'the need to maintain land-based practices as critical components of tribal identities continues to make the topic of land reacquisition and consolidation central to the study Indigenous environmental issues, and, despite its conceptual flaws, Indigenous sovereignty is a critical tool in this process'.[23]

Posthumanism and Indigenous Ontologies

The recent ontological turn has implications for questions of how ideas of land are constituted and what that means for Indigenous struggles for land and decolonization. A number of scholars have argued that by overlooking Indigenous ontologies, posthumanism misses critical insights that might be gained from Indigenous perspectives. Kim TallBear argues that 'indigenous standpoints accord greater animacy to nonhumans, including nonorganisms, such as stones and places, which help form (Indigenous) peoples as humans constituted in complex ways than in simple biological terms'.[24] She argues, along with others, that Indigenous peoples have more intimate and complex sets of relations with the animate/inanimate agents bound up in land than much posthumanist scholarship can capture. What is seen as

21   Clint Carroll, *Roots of our Renewal: Ethnobotany and Cherokee Environmental Governance* (Minneapolis, MN: University of Minnesota Press, 2015), p. 8, https://doi.org/10.5749/minnesota/9780816690893.001.0001

22   Taiaiake Alfred, *Peace, Power, Righteousness: An Indigenous Manifesto* (Toronto: Oxford University Press, 1999); Paul Nadasdy, *Sovereignty's Entailments: First Nation State Formation in the Yukon* (Toronto: University of Toronto Press, 2017), https://doi.org/10.3138/9781487515720

23   Carroll, 'Native Enclosures', 38.

24   Kim Tallbear, 'Beyond the Life/Not-life Binary: A Feminist-Indigenous Reading of Cryopreservation, Interspecies Thinking, and the New Materialisms', in *Cryopolitics: Frozen Life in a Melting World*, ed. by Joanna Radin and Emma Kowal (Cambridge, MA: MIT Press, 2017), pp. 179–201, at 187, https://doi.org/10.7551/mitpress/10456.003.0015

'alive' in much posthumanist discourse is more limited than in much Indigenous thinking. Zoe Todd makes a case for the need to decolonize posthumanist scholarship and questions the locus of agency ascribed in Eurocentric thinking.[25] Juanita Sundberg similarly argues that 'Anglo-European scholarship is the only tradition truly alive in posthumanist theorizing'.[26] She argues that all other scholarship or epistemologies are treated as truly dead through their exclusion. Understanding that the human/nature divide is far from universal is key for decolonizing and expanding posthumanist scholarship for these scholars. Many offer Indigenous ontologies, asserting that they capture more nuanced perspectives than a simple erasure of a nature/culture schism can capture. Vanessa Watts, for example offers Place-Thought as a way of framing an understanding that land is alive and thinking and, 'that humans and non-humans derive agency through the extensions of these thoughts'.[27] She frames the world in this view as a space 'where place and thought were never separated because they never could or can be separated'.[28] This framing helps overcome what she sees as the problem of subjugated agency for nonhumans in posthumanist scholarship where 'the controversial element of agency is often redesigned when applied to non-humans, thereby keeping this epistemological-ontological divide intact'.[29] Sundberg, drawing from both Sami scholar Rauna Kuokkanen and the Zapatistas' framing of the pluriverse, highlights the importance of 'multiepistemic literacy' in an expansive posthumanism that doesn't subordinate particular ontologies and forms of agency.[30] These Indigenous ontologies of land are oriented around relationality and reciprocal obligations among humans and the other-than-human. For these scholars, land, as a relationship consisting of complex and non-subjugated agencies, is key to overcoming the ontological hurdles of Eurocentric imaginings of posthumanism that these authors critique.

---

25   Zoe Todd, 'Indigenizing the Anthropocene', in *Art in the Anthropocene: Encounters Among Aesthetics, Politics, Environments and Epistemologies*, ed. by Heather Davis and Etienne Turpin (London: Open Humanities Press, 2015), pp. 241–54.

26   Juanita Sundberg, 'Decolonizing Posthumanist Geographies', *Cultural Geographies*, 21.1 (2014), 33–47, at 38, https://doi.org/10.1177/1474474013486067

27   Vanessa Watts, 'Indigenous Place-Thought and Agency amongst Humans and Non Humans (First Woman and Sky Woman go on a European World Tour!)', *Decolonization: Indigeneity, Education & Society*, 2.1 (2013), 20–34, at 21.

28   Watts, 'Indigenous Place-Thought', 21.

29   Ibid., 29.

30   Sundberg, 'Decolonizing Posthumanist Geographies'.

Brief Concluding Thoughts

Land is not just a material object, but a 'way of knowing, of experiencing and relating to the world and with others'.[31] Drawing upon work in Indigenous studies, posthumanism, and political ecology that highlights the importance of relationality and reciprocation across the human and other-than-human, we are poised to better address the politics at stake in ontologies of land by attending to the possibilities of hybridity. Structures of dispossession are not only defined by their economic or political valence to settler society, but through the notions, practices, and representations they obfuscate. Indigenous movements for social and environmental justice are deeply tied to issues of land rights. By operating on multiple ontological registers rather than the occlusion of one mode by another, Indigenous movements can focus not only on asserting ownership over lands, but on revitalizing the land-based practices that shape the fundamental nature of relationality.

Bibliography

Agrawal, Arun, *Environmentality: Technologies of Government and the Making of Subjects* (Durham, NC: Duke University Press, 2005), https://doi.org/10.1215/9780822386421

Alfred, Taiaiake, *Peace, Power, Righteousness: An Indigenous Manifesto* (Toronto: Oxford University Press, 1999).

Blaikie, Piers, *The Political Economy of Soil Erosion in Developing Countries* (London: Longman, 1985).

Blaikie, Piers and Harold Brookfield, *Land Degradation and Society* (London: Methuen, 1987).

Braun, Bruce, 'Producing Vertical Territory: Geology and Governmentality in Late Victorian Canada', *Cultural Geographies*, 7.1 (2000), 7–46, https://doi.org/10.1177/096746080000700102

Burow, Paul Berne, Samara Brock, and Michael R. Dove, 'Unsettling the Land: Indigeneity, Ontology, and Hybridity in Settler Colonialism', *Environment and Society: Advances in Research*, 9 (2018), 57–74, https://doi.org/10.3167/ares.2018.090105

Campbell, Jeremy M., *Conjuring Property: Speculation and Environmental Futures in the Brazilian Amazon* (Seattle, WA: University of Washington Press, 2015).

---

31 Coulthard, *Red Skin, White Masks*, p. 61.

Carroll, Clint, 'Native Enclosures: Tribal National Parks and the Progressive Politics of Environmental Stewardship in Indian Country', *Geoforum*, 53 (2014), 31–40, https://doi.org/10.1016/j.geoforum.2014.02.003

— *Roots of our Renewal: Ethnobotany and Cherokee Environmental Governance* (Minneapolis, MN: University of Minnesota Press, 2015), https://doi. org/10.5749/minnesota/9780816690893.001.0001

Coulthard, Glen, *Red Skin, White Masks: Rejecting the Colonial Politics of Recognition* (Minneapolis, MN: University of Minnesota Press, 2014), https://doi. org/10.5749/minnesota/9780816679645.001.0001

Ingold, Tim, *The Perception of the Environment: Essays on Livelihood, Dwelling, and Skill* (London: Routledge, 2000), https://doi.org/10.4324/9780203466025

Latour, Bruno, *Science in Action: How to Follow Scientists and Engineers through Society* (Cambridge, MA: Harvard University Press, 1987).

Li, Tania Murray, *Land's End: Capitalist Relations on an Indigenous Frontier* (Durham, NC: Duke University Press, 2014), https://doi.org/10.1215/9780822376460

— 'What is Land? Assembling a Resource for Global Investment', *Transactions of the Institute of British Geographers*, 39.4 (2014), 589–602, https://doi. org/10.1111/tran.12065

Luke, Timothy W., 'Developing Planetarian Accountancy: Fabricating Nature as Stock, Service, and System for Green Governmentality', in *Nature, Knowledge and Negation*, ed. by Harry Dahms (Bingley, UK: Emerald Group Publishing Limited, 2009), pp. 129–59, https://doi.org/10.1108/ S0278-1204(2009)0000026008

Middleton, Beth Rose, 'Jahát Jatítotòdom: Toward an Indigenous Political Ecology', in *The International Handbook of Political Ecology*, ed. by Raymond Bryant (Northampton, MA: Edward Elgar Publishing, 2015), pp. 561–76, https://doi.org/10.4337/9780857936172.00051

Mitchell, Timothy, *Colonising Egypt* (Berkeley, CA: University of California Press, 1991).

— *Rule of Experts: Egypt, Techno-Politics, Modernity* (Berkeley, CA: University of California Press, 2002).

Moore, Donald S., *Suffering for Territory: Race, Place, and Power in Zimbabwe* (Durham, NC: Duke University Press, 2005), https://doi. org/10.1215/9780822387329

Nadasdy, Paul, 'The Gift in the Animal: The Ontology of Hunting and Human-Animal Sociality', *American Ethnologist*, 34.1 (2007), 25–43, https://doi. org/10.1525/ae.2007.34.1.25

— *Hunters and Bureaucrats: Power, Knowledge, and Aboriginal-State Relations in the Southwest Yukon* (Vancouver: UBC Press, 2003).

— *Sovereignty's Entailments: First Nation State Formation in the Yukon* (Toronto: University of Toronto Press, 2017), https://doi.org/10.3138/9781487515720

Robbins, Paul, *Political Ecology: A Critical Introduction* (New York, NY: John Wiley & Sons, 2012).

Scott, James C., *Seeing Like a State: How Certain Schemes to Improve the Human Condition Have Failed* (New Haven, CT: Yale University Press, 1998).

Sundberg, Juanita, 'Decolonizing Posthumanist Geographies', *Cultural Geographies*, 21.1 (2014), 33–47, https://doi.org/10.1177/1474474013486067

TallBear, Kim, 'Beyond the Life/Not-life Binary: A Feminist-Indigenous Reading of Cryopreservation, Interspecies Thinking, and the New Materialisms', in *Cryopolitics: Frozen Life in a Melting World*, ed. by Joanna Radin and Emma Kowal (Cambridge, MA: MIT Press, 2017), pp. 179–201, https://doi.org/10.7551/mitpress/10456.003.0015

Todd, Zoe, 'Indigenizing the Anthropocene', in *Art in the Anthropocene: Encounters Among Aesthetics, Politics, Environments and Epistemologies*, ed. by Heather Davis and Etienne Turpin (London: Open Humanities Press, 2015), pp. 241–54.

Watts, Vanessa, 'Indigenous Place-Thought and Agency amongst Humans and Non Humans (First Woman and Sky Woman Go on a European World Tour!)', *Decolonization: Indigeneity, Education & Society*, 2.1 (2013), 20–34.

SECTION VI

THE WEAVE OF EARTH AND COSMOS

Fig. 9 Pedra Azul Milky Way. Photo by Eduardo M. S. Neves (2017), Wikimedia, CC BY SA 4.0, https://commons.wikimedia.org/wiki/File:Pedra_Azul_Milky_Way.jpg#/media/File:Pedra_Azul_Milky_Way.jpg

18. Gaia and a Second Axial Age

Sean Kelly

For the first time in sixty-five million years, the Earth community is being drawn into a collective, planet-wide near-death experience (NDE). While NDEs are known to occur spontaneously, they (or their experiential or symbolic analogues) have also been intentionally cultivated in all traditional societies as an essential moment of rites of passage or initiation. In all such rites, the initiation and its confrontation with death are not random events, but intentional processes, guided or lured by a specific goal. They are, in technical terms, teleological in nature (from *telos*: purpose, goal). For individual initiations, the purpose or goal involves the emergence of a new identity (whether of shaman, healer, chief, warrior etc.), an identity molded to serve the interests of the wider community. Something similar is happening in our times with regard to the threshold on which the Earth community now finds itself poised. In this case, however, we are dealing with the emergence of a radically new kind of identity or subject on a planetary scale. I use the word *Gaian* here as it seems, more than any other I have encountered, to be a kind of strange attractor for many of the more creative manifestations of the emerging planetary identity.

Though accelerating in our own times, the emergence of this new identity has in fact long been in the making. Over two centuries ago now, G. W. F. Hegel announced that 'ours is a birth-time and a period of transition to a new era. Spirit has broken with the world it has hitherto inhabited and imagined, and is of a mind to submerge it in the past, and in the labor of its own transformation...'.[1] We are still in this period of transition, but the pace has quickened and the stakes are higher than

---

1   G. W. F. Hegel, *Phenomenology of Spirit*, trans. by A. V. Miller (New York, NY: Oxford University Press, 1977), p. 6.

    https://doi.org/10.11647/OBP.0186.18

Hegel could have imagined: not only the five-thousand year old fabric of civilization, but that of life itself on a planetary scale. With runaway climate change and the mass extinction of species well underway, we can legitimately say that we live in end times (*eschaton*). Whether or not, on the other side of the eschaton, a better — or at least a viable and potentially flourishing — world awaits us, depends upon how we navigate the next decade or so.

Hegel was not the first to announce the dawning of a new age. He stood quite consciously in a long line of prophet-seers, from his immediate Enlightenment predecessors (Kant, Condorcet), through the mystic Jacob Boehme and the pivotal figure of Joachim di Fiore, all the way to St. Paul and the 'veil-lifting' (*apocalypsis*) claims of the New Testament. I have argued elsewhere that the birth and ongoing transformation of the modern period, or Planetary Era, is prefigured in certain central Biblical symbols which act, in William Blake's expression, as the 'Great Code' not only of art and literature, but of the deeper patterning of world history.[2] Whether or not one agrees with my argument, one can perhaps grant the deep resonance between our planetary moment and the New Testament's heightened sense of living in end times with a longing for a New Age.

Though falling outside its assigned limits, the Christianity of the New Testament is nevertheless a hybrid product of the earlier great transformation that Karl Jaspers termed the Axial Age. So much of what was to become the foundations of the world's great religions, major philosophies, and dominant worldviews were laid down from the eighth to the third centuries BCE, the period Jaspers assigned to the Axial Age. With the near simultaneous emergence around the sixth century BCE of the first Greek philosophers (from Thales and Pythagoras to Plato), the Buddha, Mahavira, Confucius, and Lao Tzu, the great Jewish prophets (Second Isaiah, Ezekiel, Jeremiah), and possibly Zoroaster, this period 'gave birth to everything which, since then, man has been able to be, the point most overwhelmingly fruitful in fashioning humanity'. It is during this period 'that we meet with the most deepcut dividing line in history. Man, as we know him today, came into being'.[3] If Jaspers were

2   Sean Kelly, *Coming Home: The Birth and Transformation of the Planetary Era* (Great Barrington, MA: Lindisfarne Books, 2010).

3   Karl Jaspers, *The Origin and Goal of History*, trans. by Michael Bullock (New York, NY: Routledge, 2014), p. 1, https://doi.org/10.4324/9781315823683

alive today, he might see our own times as straddling 'the most deepcut dividing line in history'. He might as well have come to believe, as many do today, that we are witness to the dawning of a Second Axial Age.

While the idea of a Second Axial Age seems originally to have been proposed by Thomas Berry, the first extended treatment in print was by Ewert Cousins, who summarized his understanding as follows:

> Having developed self-reflective, analytic, critical consciousness in the First Axial Period, we must now, while retaining these values, reappropriate and integrate into that consciousness the collective and cosmic dimensions of the pre-Axial consciousness. We must recapture the unity of tribal consciousness by seeing humanity as a single tribe. And we must see this single tribe related organically to the total cosmos. This means that the consciousness of the twenty-first-century will be global from two perspectives: (1) from a horizontal perspective, cultures and religions are meeting each other on the surface of the globe, entering into creative encounters that will produce a complexified collective consciousness; (2) from a vertical perspective, they must plunge their roots deep into the earth in order to provide a stable and secure base for future development. This new global consciousness must be organically ecological, supported by structures that will ensure justice and peace. In the Second Axial Period this twofold global consciousness is not only a creative possibility to enhance the twenty-first century; it is an absolute necessity if we are to survive.[4]

The dominant strands of first Axial traditions tended to emphasize the transcendent pole in the vertical dimension (as we see in Platonic and the later Cartesian dualisms; Christian otherworldliness; Hindu and Buddhist views of the 'wheel of life' as illusion or trap; in Chinese cosmology, the immovable Pole star as symbol of Heavenly power and virtue, or the Daoist immortals). In the extreme, according to Robert Bellah, these dominant strands involved 'the religious rejection of the world characterized by an extremely negative evaluation of man and society and the exaltation of another realm of reality as alone true and infinitely valuable'.[5] At the same time, while the first Axial Age involved a new consciousness of the universal in its noetic, cosmic,

---

4   Ewert Cousins, *Christ of the 21ˢᵗ Century* (Rockport, MA: Element Books, 1992), p. 10, https://doi.org/10.5040/9781472550200

5   Robert Bellah and Hans Joas, eds, *The Axial Age and its Consequences* (Cambridge, MA: Harvard University Press, 2012), p. 194, https://doi.org/10.4159/harvard.9780674067400

and ethical dimensions, the several axial epiphanies of the universal remained rooted in the exclusive (ethno-linguistic) particularities of their respective culture spheres, and therefore in both of these senses the universal was abstract. 'Great as the major figures of the axial age were', as Bellah would note in his last and greatest work,

> and universalistic as their ethics tended to be, we cannot forget that each of them considered his own teaching to be the only truth or the highest truth, even such a figure as the Buddha, who never denounced his rivals but only subtly satirized them. Plato, Confucius, Second Isaiah, all thought that it was they and they alone who had found the final truth. This we can understand as an inevitable feature of the world so long ago.[6]

A central task of the Second Axial Age, by contrast, involves the articulation of new forms of universality which could mediate between the particular culture spheres and help them confront their shared predicament: the threat of planet-wide ecological and civilizational collapse.

Despite the astounding synchronicity of first Axial Age, it was not global or planetary in extent, and its various representatives were largely unconscious of the parallel developments outside of their own culture spheres. At the same time, however — and as I have argued in detail in *Coming Home* — it was the destiny of one late, hybrid, shoot (Christianity) of this first axial mutation to become the symbolic catalyst or lure for the eventual emergence, some fifteen hundred years later, of the Planetary Era (more commonly designated as the modern period). It is with this specific genealogical line that we can discern an answer to the question of the relation between the two Axial Ages: the central symbols of Incarnation (of Spirit into matter, of the Logos into Cosmos, of the eternal into time) and of God as Trinity (the Absolute as internally differentiated) prefigure the deep structure of the movement from the first to the second Axial Age, with Modernity as the middle term between both Ages. The first Axial Age sets up the conditions of possibility for the eventual emergence of the second. These conditions include the reflexive and critical consciousness associated with 'metacognition/theoretic culture' (M. Donald); the 'disembedding'

---

6   Robert Bellah, *Religion in Human Evolution: From the Paleolithic to the Axial Age* (Cambridge, MA: Harvard University Press, 2011), p. 602, https://doi.org/10.4159/ harvard.9780674063099

(Charles Taylor) of culture from the cosmos and of the individual from the collective; and the lure of the universal (Eric Voegelin, Jan Assmann).[7] At a deeper level, both ages can and should be seen as the two poles of a single process, or rotating axis, moving from the abstract to the concrete in three broad phases: (i) an initial identity (in this case, structured around the central myth/symbol of Incarnation wedded to the Greek intuition of the universal as logos/cosmos); (ii) a movement of differentiation — and later, dissociation (leading to the birth of modern science, the modern disengaged subject, and the broader processes of secularization; all of which bring about the birth of the Planetary Era and the accelerating planetary crisis); (iii) a new Gaian, or *Gaianthropic* identity in the making.[8] The cultivation of this new identity is a central task of the Second Axial Age, which itself can be seen as the 'opportune moment' (*kairos*) for the actualization of the deeper telos of the 2,500-year Axial Aion. My proposal for the periodization of the larger arc that encompasses the two Axial Ages is as follows.

Axial Aion (c. 800BCE to present):
First Axial Age (c. 800–200BCE)
Planetary Era (c. 1500CE to present)
Second Axial Age (Gaian epoch or Gaianthropocene) (c. 1945?–)

In contrast to the abstract universals that dominated the first Axial traditions, the new Gaian identity exemplifies the real-ideal of concrete universality. The universality of Gaia consists most obviously in the fact that it is in and through Gaia that we live and have our being. Gaia is the *ground* of what we all share in *common*. For the same reason, this universality is concrete, to begin with, in the sense that the physical systems studied by Earth System science constitute the *shared*, living

---

7   M. Donald, 'An Evolutionary Approach to Culture: Implications for the Study of the Axial Age', in *The Axial Age and its Consequences*, pp. 47–76, https://doi.org/10.4159/harvard.9780674067400.c3; Charles Taylor, 'What was the Axial Revolution?', in *The Axial Age and its Consequences*, pp. 30–46, https://doi.org/10.4159/harvard.9780674067400.c2; Eric Voegelin, *Order and History, Vol. 4, The Ecumenic Age* (Columbia, MO: University of Missouri Press, 1974); Jan Assmann, 'Cultural Memory and the Myth of the Axial Age', in *The Axial Age and its Consequences*, pp. 366–408, https://doi.org/10.4159/harvard.9780674067400.c15

8   See Sean Kelly, 'Cosmological Wisdom and Planetary Madness', *Tikkun Magazine*, 11 November 2015, https://www.tikkun.org/nextgen/cosmological-wisdom-planetary-madness, for my introduction of this term.

body of the entire Earth community. It is also concrete, however, in the specifically Hegelian sense that it, or She, is *auto*-poietic or *self*-organizing (the foundational insight of James Lovelock and Lynn Margulis' Gaia Theory), which is to say that Gaia is a *Subject* (as well as a communion of subjects, and not a mere collection of objects).

The actualization of concrete universality that I see as the guiding spirit of the Second Axial Age will depend upon the successful coordination of multiple initiatives, both theoretical and practical, across the full spectrum of human endeavor. Here I will focus on some key features of the theoretical. If the First Axial Age was associated with the emergence of theoretic culture, with its second-order thinking, metacognition, and radical mytho-speculation, the Second Axial Age is marked by what could be described as third-order metacognition and a new (planetary) radical mytho-speculation. Integrating the critical, reflexive virtues of first-Axial theoretic culture, the leading edge of theory in the Second Axial Age recognizes the destructive potential of the disembedded, disengaged subject (which reduces the world to a mere collection of objects). It re-embeds the human subject into the living Earth and cosmos — or rather renews consciousness of the fact, and mystery, of its ontological consubstantiality with Earth and cosmos — which are now seen, celebrated, and engaged with as a communion of subjects. More radically, we can say that the radiating center of the second Axial Age is constituted by an awareness — a third-order metacognition — in a growing network of individuals and communities, that 'We live in that time when Earth itself begins its adventure of conscious self-awareness'.[9]

This awareness is informed and catalyzed by many distinct, if overlapping, disciplines, including Earth System science, Big History, the various strands of ecological science and environmental studies, the field of religion and ecology, and the emerging transdiscipline of integral ecology. For our purposes, I would single out the generative contributions of Thomas Berry, one of the founders of integral ecology and, along with Brian Swimme, Mary Evelyn Tucker, and others, bard or prophet of a more coherent and inspiring Big, or better, Deep History. 'We need to think of the planet', Berry writes, 'as a single, unique,

---

9   Brian Swimme and Mary Evelyn Tucker, *Journey of the Universe* (New Haven, CT: Yale University Press, 2011), p. 109.

articulated subject to be understood in a story both scientific and mythic [and, I would add, ethical and political]'.[10]

In the variety of Big History associated with David Christian, Earth or Gaian evolution is conceived as currently poised on 'Threshold 9' (the previous eight thresholds are: (1) origin of universe; (2) formation of stars and galaxies; (3) formation of heavier chemical elements; (4) formation of Earth and the solar system; (5) emergence of life; (6) birth of *homo sapiens*; (7) agricultural revolution; (8) the modern revolution, or what I call the Planetary Era). Christian has little to say about this new evolutionary threshold, other than underlining its radically uncertain character (which in any case attaches to the emergent properties associated with all new thresholds) and the possibility, at least, of somehow achieving a sustainable planetary civilization. Futurist and Big Historian Joseph Voros concludes soberly that the most likely path ahead involves 'a slowly-unfolding collapse or "descent" over a time-scale of decades-to-centuries towards a human society characterized by ever-declining access to sources of fossil fuel-based energy'.[11] At the same time, however, drawing from fellow futurist James Dator's fourfold typology of alternative futures, Voros leaves open the possibility of Threshold 9 involving an eventual transition to a planetary 'transformational society', visions of which tend to emphasize either technological breakthroughs or the actualization of spiritual potentials (the other three possible futures are: continued growth, collapse, and disciplined society).[12] In the latter case, 'some new form or aspect of human consciousness emerges and redefines our value systems, such that we become focused on "higher" goals than we currently pursue'. It might be argued, he continues, 'that Cosmic Evolution, Big History and other related conceptual frameworks may themselves provide a foundation for a new more integrated worldview, onto which an almost spiritual dimension could

---

10 Thomas Berry, *The Sacred Universe: Earth, Spirituality, and Religion in the Twenty-First Century* (New York, NY: Columbia University Press, 2009) p. 112.

11 James Voros, 'Profiling "Threshold 9": Using Big History as a Framework for Thinking about the Contours of the Coming Global Future', in *Evolution: Development within Big History, Evolutionary and World-System Paradigms*, ed. by Leonid E. Grinin and Andrey V. Korotayev (Volgograd: Uchitel Publishing House, 2013), pp. 119–42, at p. 119.

12 See C. Bezold, 'Jim Dator's Alternative Futures and the Path to IAF's Aspirational Futures', *Journal of Futures Studies*, 14.2 (2009), 123–34.

be read'.[13] Clearly what I am proposing in terms of the emergence of a new Gaian or Gaianthropic identity qualifies as such a new, radical transformation of consciousness.

Returning to the theme of initiation that I opened with, we can note that an essential component during the liminal ('threshold') phase of many rites of initiation involves introducing the initiate to the sacred stories, myths, and symbols of the community into which they are being inducted. In contrast to the situation in both archaic or Indigenous societies and in first Axial traditions, the stories and symbols required for this collective initiation into a new Gaian identity need to include a genuinely common narrative core, regardless of language and ethnicity. The only candidate in this case is the Universe story itself, the major lines, phases, and thresholds of which are, for the first time in the history of our species, well understood and universally acknowledged by the many scientific communities devoted to their study. This is not to say that there is no longer a place for the stories, myths, rituals, and doctrines of the world religions and of Indigenous cultures. On the contrary, these should continue to provide inspiration for reflection on the mysteries of the cosmos, of human nature, and the question of spiritual ultimates, including indications of how we might best navigate the critical planetary threshold on which we are so precariously poised. It is precisely to this end that so much fine work is now being done in the field of religion and ecology.[14]

Of course, we cannot know, or at least we cannot expect to arrive at a general consensus as to whether or not one or the other of the world's religious traditions might actually have a direct line to the Universal (or Universals) intuited during the first Axial Age. We can and must, however, acknowledge the sacred character of Gaia as concrete universal. The story of Gaia is sacred because it tells of our common origin and will include our shared destiny. We can therefore envision the open spaces of this living Earth, in and through whom we literally have our being, as the Common Temple of the Second Axial Age, with the sacred places explored by the world's religious traditions as so many side temples with their own unique paths leading to the great Mystery.

---

13   Voros, 'Profiling "Threshold 9"'.
14   John Grim and Mary Evelyn Tucker, *Ecology and Religion* (Washington, DC: Island Press, 2014).

It is possible that a new, third kind or species of religion might emerge from our gathering in the sacred precincts of this Common Temple. As Edgar Morin puts it, this would be religion 'in the minimal sense' of the term (suggested in one derivation of the word: from *re-ligare*: to join back together), at the heart of which would be the fact and ideal of planetary 're-liance'.[15] While the first kind of religion arose out the first Axial Age, and the second kind in the Modern period with its faith in this-worldly salvation (the myth of Reason, of progress, and 'development'), the new religion, by contrast,

> would not have promises but roots: roots in our cultures and civilizations, in planetary and human history; roots in life; roots in the stars that have forged the atoms of which we are made; roots in the cosmos where the particles were born and out of which our atoms were made [...] Such a religion would involve belief, like all religions but, unlike other religions that repress doubt through excessive zeal, it would make room for doubt within itself. It would look out onto the abyss.[16]

The Earth community is being dragged to the edge of this abyss. Faced with the prospect of ever more probable civilizational collapse and an accelerating mass extinction, the human members of this community must learn to think, feel, and act out of their wider and deeper identity as Gaia. There is no guarantee that we will avert planetary catastrophe. There never has been such a guarantee. We can, however, still accomplish the task that has been the secret preoccupation of the 2,500-year Axial Aeon, if not of the 4.6-billion-year journey of Earth's evolution. In the meantime, in the words of the poet,

> Let everything happen to you: beauty and terror.
> Just keep going. No feeling is final.
>
> Don't let yourself lose me.
>
> Nearby is the country they call life.
> You will know it by its seriousness.[17]

---

15 Edgar Morin, with Brigitte Kerne, *Homeland Earth: A Manifesto for the New Millennium* (New York, NY: Hampton Press, 1999), p. 141.

16 Ibid., p. 142.

17 Rainer Maria Rilke, *Book of Hours*, trans. by Anita Barrows and Joana Macy (New York, NY: Riverhead Books, 1996), I, 59.

214 *Living Earth Community*

Bibliography

Assmann, Jan, 'Cultural Memory and the Myth of the Axial Age', in *The Axial Age and its Consequences*, ed. Robert Bellah and Hans Joas (Cambridge, MA: Harvard University Press, 2012), pp. 366–408, https://doi.org/10.4159/harvard.9780674067400.c15

Bellah, Robert, *Religion in Human Evolution: From the Paleolithic to the Axial Age* (Cambridge, MA: Harvard University Press, 2011), https://doi.org/10.4159/harvard.9780674063099

Bellah, Robert and Hans Joas, eds, *The Axial Age and its Consequences* (Cambridge, MA: Harvard University Press, 2012), https://doi.org/10.4159/harvard.9780674067400

Berry, Thomas, *The Sacred Universe: Earth, Spirituality, and Religion in the Twenty-First Century* (New York, NY: Columbia University Press, 2009).

Bezold, C., 'Jim Dator's Alternative Futures and the Path to IAF's Aspirational Futures', *Journal of Futures Studies*, 14.2 (2009), 123–34.

Cousins, Ewert, *Christ of the 21ˢᵗ Century* (Rockport, MA: Element Books, 1992), https://doi.org/10.5040/9781472550200

Donald, M., 'An Evolutionary Approach to Culture: Implications for the Study of the Axial Age', in *The Axial Age and its Consequences*, ed. by Robert Bellah and Hans Joas (Cambridge, MA: Harvard University Press, 2012), pp. 47–76, https://doi.org/10.4159/harvard.9780674067400.c3

Gebser, Jean, *The Ever-Present Origin* (Athens, OH: Ohio University Press, 1986).

Grim, John, and Mary Evelyn Tucker, *Ecology and Religion* (Washington, DC: Island Press, 2014).

Hegel, G. W. F., *The Phenomenology of Spirit*, trans. by A. V. Miller (New York, NY: Oxford University Press, 1977).

Jaspers, Karl, *The Origin and Goal of History*, trans. by Michael Bullock (New York, NY: Routledge, 2014), https://doi.org/10.4324/9781315823683

Kelly, Sean, *Coming Home: The Birth and Transformation of the Planetary Era* (Great Barrington, MA: Lindisfarne Books, 2010).

— 'Cosmological Wisdom and Planetary Madness', *Tikkun Magazine*, 11 November 2015, https://www.tikkun.org/nextgen/cosmological-wisdom-planetary-madness

Lovelock, James, *Gaia: A New Look at Life on Earth* (New York, NY: Oxford University Press, 2016).

Morin, Edgar, with Brigitte Kerne, *Homeland Earth: A Manifesto for the New Millennium* (New York, NY: Hampton Press, 1999).

Rilke, Rainer Maria, *Book of Hours*, trans. by Anita Barrows and Joana Macy (New York, NY: Riverhead Books, 1996).

Swimme, Brian, and Mary Evelyn Tucker, *Journey of the Universe* (New Haven, CT: Yale University Press, 2011).

Swimme, Brian, and Thomas Berry, *The Universe Story: From the Primordial Flaring Forth to the Ecozoic Era — A Celebration of the Unfolding of the Cosmos* (San Francisco, CA: HarperOne, 1992).

Taylor, Charles, 'What Was the Axial Revolution?', in *The Axial Age and its Consequences*, ed. by Robert Bellah and Hans Joas (Cambridge, MA: Harvard University Press, 2012), pp. 30–46, https://doi.org/10.4159/harvard.9780674067400.c2

Voegelin, Eric, *Order and History, Vol. 4, The Ecumenic Age* (Columbia, MO: University of Missouri Press, 1974).

Voros, James, 'Profiling "Threshold 9": Using Big History as a Framework for Thinking about the Contours of the Coming Global Future', in *Evolution: Development within Big History, Evolutionary and World-System Paradigms*, ed. by Leonid E. Grinin and Andrey V. Korotayev (Volgograd: Uchitel Publishing House, 2013), pp. 119–42.

19. The Human Quest to Live in a Cosmos[1]

Heather Eaton

The cosmos is also within us. We are made of star-stuff. We are a way for the universe to know itself. (Carl Sagan).[2]

Introduction

Throughout human history, there is a steady, even unrelenting, lure to understand the facets of the 'orders of reality'. Humans look outward, to touch and gauge the limits of reality, to perceive its ethos and telos. The quest to observe and fathom the exterior edges of reality has been central to religious, scientific, and philosophical inquiry and meaning making over millennia. The quest to live in a cosmos is more than knowledge: it is an orientation to living within Earth now. It represents many journeys: an outward journey to the boundaries of the universe, Earth's journey, the human journey, and the interior journey of integrating these together.

a) Exterior Quest: Where Are We?

The expansiveness and entanglement of time, space, and materiality are mesmerizing actualities. There is a steady, even unrelenting, lure to apprehend the facets of the 'orders of reality', that enable us to see

---

1   This is an abridged version of a chapter of the same title in *Encountering Earth: Thinking Theologically with a More-Than-Human World*, ed. by Trevor Bechtel, Matt Eaton, and Tim Harvie (Eugene, OR: Wipf and Stock Publishers, 2018), pp. 227–47.

2   'The Shores of the Cosmic Ocean [Episode 1]', *Cosmos: A Personal Voyage*, PBS, 28 September 1980.

          https://doi.org/10.11647/OBP.0186.19

or perceive *where* we are. For Thomas Berry, the expansiveness and essence of the world — cosmos, Earth, time, space, and processes — are central to knowing anything meaningful about being human.[3] In order to respond to the ecological crisis, which is planetary, it is necessary to know the history, origins, and dynamics of planet Earth. Earth has its origins in the dynamics and processes of the universe.

The universe — the farthest realm of the exterior quest — is understood with increasing clarity. Scientific modes of inquiry progressively detect the dynamic processes, interconnections, and expansions of the universe. What is increasingly astonishing is that everything about the universe is so much more than assumed or imagined previously. Evidence abounds about the complexities, diversifications, and the development sequences of transformations, and the intricacies and inter-relatedness of the emergent universe. It is also increasingly apparent that the universe is integral: unified without being uniform. There is a cohesiveness within the astonishing diversity found in how the universe functions, including in the birth and death of stars, and galaxy and planetary formations.

Overall, in these vast exterior realms of reality, there are patterns, processes, developmental sequences, transformations, evolutions, intensifications and complexifications. For example, in the transformation from the atomic to the molecular structures, a further degree of intensity develops in these new physical arrangements. One could say reality complexifies. Furthermore, it is entangled. Although each discipline explains particular processes in discrete separated categories, if we step back, it is evident that they are interconnected and interdependent processes. How could it be otherwise? Thus, scientists are using terms such as emergent complexity, entanglement, coherence, correspondence, congruence, or intelligibility to describe the overall coordination within the universe.[4]

Coherence and integration are also seen in the evolution and functioning of the biosphere. The biosphere is best described, and

---

3   Thomas Berry, *Dream of The Earth* (San Francisco, CA: Sierra Club Books, 1988). For a full bibliography of Thomas Berry's works see http://thomasberry.org/life-and-thought/bibliography. See also *The Intellectual Journey of Thomas Berry: Imagining the Earth Community*, ed. by Heather Eaton (Lanham, MD: Lexington Press, 2014).

4   Karen Barad, *Meeting the Universe Half-Way: Quantum Physics and the Entanglement of Matter and Meaning* (Durham, NC: Duke University Press, 2007), https://doi.org/10.1215/9780822388128

explained, with inter-related processes, networks of connections, correspondence, mutual influences, and communication from the molecular and cellular to the planetary processes. More is known about the boundaries of exterior realities than ever before. One outcome of making this knowledge public and accessible is to amplify and intensify human consciousness about 'where we are'. Thus, humans are becoming more conscious of the universe and Earth. For Berry, there is an urgent need to consider how to interpret this new knowledge of the universe.

The cosmos cannot be seen as a backdrop to the human drama, or as a context, an unfolding, a progression or a potential. It is not like an embryo that matures into fullness. It is more a becoming: not linear and determined, but creative and dynamic, yet, seemingly, with an orientation. As the universe develops, it becomes more: more complex, interactive, entwined, vibrant, and intense. That is why, for Pierre Teilhard de Chardin and Berry, the best image is that the cosmos is a cosmogenesis. This implies forms of continuity and coherence between cosmogenesis, geogenesis, and biogenesis. In the same manner, evolution is a process or dynamic of the biosphere from which *homo sapiens* evolved, with a form of self-reflexive symbolic consciousness that is able to perceive that these forms of genesis are ongoing. There is coherence and continuity.[5]

What Berry realized is that this is radically new knowledge. The reference points for understanding the universe, Earth, ourselves, and our role within the scheme of things all change with this new knowledge. To understand anything, we need to grasp, even at a basic level, that the universe is a primary source and reference. Everything about Earth evolved and developed from cosmic processes. All aspects of *homo sapiens* evolved and developed from Earth processes. To say that Earth formed or produced us is inadequate language. We emerged from and are a conscious living part of Earth realities. By extension and extrapolation, the most apt description of the universe is that it is alive.[6]

In this vein, the expansion of human consciousness into the cosmos is also the universe and Earth becoming conscious in humanity. Put

---

5   This is not uniformity, or intelligent design where the configuration was predestined.
6   A magnificent exposé of this understanding is in Mary Evelyn Tucker and John Grim's edited volume, *Living Cosmology: Christian Responses to the Journey of the Universe* (Maryknoll, NY: Orbis, 2016).

differently, it is the universe reflecting on itself in human form, or that humans are a mode of self-consciousness of the universe. To understand and integrate that we are a self-conscious element of a living cosmos is a great challenge. The quest to live in a cosmos is thus a dynamic of the cosmos, and encounter with the cosmos. One way is to enter into this interpretive zone is to see how this exterior quest to know the cosmos is also an interior quest.

b) Who Are We? Exteriority Becomes Interiority

Of the myriad ways to broach this topic, the basic point is an extension of the above: the quest to understand the largest parameters of reality is intimately involved with 'who we are' in the scheme of things, and also with the interior modes of knowing. Of the countless ways to consider interiority, I have two comments related only to this chapter's theme. The first remark is about a fusion between exteriority and interiority. Here I am indebted to and influenced by significant aspects of the thought of Teilhard de Chardin. In terms of interiority, Teilhard contemplated an intimacy between the *without* and the *within of things*.[7] The without is the observable. This includes the structures and changes from the establishment and bonding arrangements of atomic structures, to the formation of molecules and mega molecules out of which arose and evolved all matter. The starting point for understanding the without is the discernable atomic structures and behaviours. Teilhard sought, and developed, a theory that connects structure and activity with processes and purposes of the developmental transformations. He pondered these as a whole, meaning he would not separate anything from its structures, activities, developments, and directionality.

For example, Teilhard studied bacteria cultures in this manner, and then plants. He explained that for plants, the without and observable cannot explain the life dynamics of plants. With insects it is yet more difficult, with vertebrates it is futile, and, with humans, it breaks down completely. As life evolved, the without of things — the observable — becomes increasingly incapable of explaining the

---

7   This concept is introduced and developed throughout Pierre Teilhard de Chardin, *The Human Phenomenon*, trans. by Sarah Appleton-Weber (East Sussex, UK: Sussex Academic Press, 1999).

behaviour, development, intensifying complexity and evolutionary directionality. Some form of interiority, 'within of things', élan vital, vitality, subjectivity, *Geist*, *qi* is increasingly present, active, effective, and, indeed, essential, everywhere.

This within of things, its interiority, is a subtle, nuanced union of matter, energy, spirit, and telos that coheres the interior dynamics with the transformations to increasing levels of complexity. The within of things is manifested in the overall orientation and processes that compel atoms to transform to molecules, to form planets, Earth, an atmosphere and biosphere, to life, consciousness, and self-consciousness. Herein we see something similar to the congruence, intelligibility, and coherence mentioned above.

Teilhard used the term consciousness with many qualifications. Consciousness itself is differentiated and evolves, from nascent traces to that of life and mind. Inert matter does not have consciousness *per se*, although through his lengthy discussions of the within of things, Teilhard maintained that something interior, not observable yet clearly present, moves the process of evolution. Overall, in *Le Phénomène Humain* [The Human Phenomenon], he wanted to write the natural history from the without and the within of things, which combined spirituality and science. This was predictably contentious for both disciplines. For me, however, it was cogent, brilliant and beautiful, and reverberated deeply with my quest.

There is a further aspect to Teilhard's thought that is important to appreciate in his elegant blend of science, religiosity, and poiesis. Each stage of evolution is nascent in the previous, but not in a simple embryonic or potential form. There are critical changes that alter the very ontology of reality. For example, in the transformation from the atomic to the molecular structures, the new arrangements of the parts required the acquisition of another dimension — a further degree of interiority — that allowed reality to complexify. Every evolutionary development requires an intensified and differentiated structure that corresponds to a more subtle and supple concentrated interiority and consciousness. Scientifically, reality thickens, deepens, and crosses new ontological thresholds. Spiritually, interiority intensifies and amplifies.

Berry concurred that there must be differentiated forms of interiority within the processes of the universe, Earth, and the biosphere.

These dynamics are the creative energies interior to all sequences of transformations. These relationships between the without and within of things, between exterior and interior dynamics and processes, occur at all times and are intensified at every development phase. Berry used the language of differentiation, subjectivity, and inter-relatedness to convey a similar insight. What is relevant here is that many of the intuitions that Teilhard proposed are being verified through science, although described with different language and not interpreted in a spiritual framework.[8]

A second remark is about interior awareness. How are we to absorb these new findings about where and who we are? Even the most hard-nosed evolutionary scientists must accept that life emerged from Earth dynamics: from the interior of cosmic and planetary processes. By extrapolation, the same must be said of consciousness, self-consciousness, and symbolic consciousness. Therefore, it is logical to claim there is an emergent interiority: highly differentiated among species. Yet, while the observations and logic are difficult to deny, the implications are far-reaching and do not fit neatly into most operative worldviews.

To study evolution is to realize that the biosphere thrives in integrated and inter-dependent relations, from the interwoven atmospheric, climate and water systems to fractal patterns and cellular dynamics. The complexity and ingenuity of Earth processes such as self-organizing dynamics, natural selection, emergence, symbiosis, and co-evolution become apparent. Earth enlivens interconnected webs of bacteria, insects, plants, animals and their related social patterns, and forms of consciousness. To attend to evolution, even minimally, is to be dazzled. Earth's intricacies animate the human imagination. The immense and elaborate planetary hydrologic cycle is stunning and breath-taking. From the microbiotic and genetic levels to the dinosaurs, the processes and life forms are astonishing. To see the elegance of birds, the ingenuity of insect communication, and the emotions of mammals is to be thrilled

---

8   For example see Lynn Margulis and Dorion Sagan, *Dazzle Gradually: Reflections on the Nature of Nature* (White River Junction, VT: Chelsea Green Publishing, 2007); Lynn Margulis, *The Symbiotic Planet: A New Look at Evolution* (New York, NY: Basic Books, 1998); Ursula Goodenough, *The Sacred Depths of Nature* (New York, NY: Oxford University Press, 2000); Elisabet Sahtouris, *Earthdance: Living Systems in Evolution* (San José: iUniverse Publication, 2000).

and overwhelmed by the creativity, diversity, power and beauty. These intimate immensities nourish human depths, or at least they could.

All animals, including humans, need first to be understood as differentiated yet integrated living elements of a whole. To grapple with the implications of evolutionary complexities propels a momentous perspectival shift. Elsewhere I have described this as the revolution of evolution.[9] It is my view, and experience, that becoming aware of the extraordinary dynamics of evolution can open up the possibility of profound depth or religious experiences. Such experiences allow a glimpse into a world of stunning elegance, of mysteries and adventure, of vistas beyond our knowing. The natural world inspires wonder and awe: a kind of power available to all who attend carefully to the natural world. The movement of the stars, the presence of mountains, the invigorating quality of ocean waves fills us with feelings of celebration and reverence. The eloquence of Abraham Joshua Heschel is worthy of a pause:

> Awe is an intuition for the dignity of all things, a realization that things not only are what they are but also stand, however remotely, for something supreme. Awe is a sense for the transcendence, for the reference everywhere to mystery beyond all things. It enables us to receive in the world intimations of the divine, [...] to sense the ultimate in the common and the simple; to feel in the rush of the passing the stillness of the eternal. What we cannot comprehend by analysis, we become aware of in awe.[10]

We are moved, like Teilhard de Chardin, to claim we live in a divine milieu and that matter, spirit, and life are intertwined in a sacred process. We can see a deeper reality: one that kindles the imagination, awakens us to the Earth, and ignites a fire and desire to protect the biosphere. Familiarity with evolution can open awareness or consciousness to Earth mysticism, a blend of the best of science and religion.[11]

---

9   Heather Eaton, 'The Revolution of Evolution', *Worldviews: Environment, Culture, Religion*, 11.1 (2007), 6–31, https://doi.org/10.1163/156853507x173478

10   Abraham Joshua Heschel, *I Asked for Wonder: A Spiritual Anthology*, ed. by Samuel Dresner (New York, NY: Crossroad, 1997), p. 3.

11   Two unrelated Kauf(f)mans have written on these topics. Stuart Kauffman, *Reinventing the Sacred: A New View of Science, Reason and Religion* (New York, NY: Perseus, 2008), and Gordon Kaufman, *In the Beginning: Creativity* (Minneapolis, MN: Fortress, 2004).

This kind of description is uncommon, and is unacceptable in many academic discourses. It is too subjective, too emotive and imprecise. It becomes a kind of ecopoiesis, and then is readily dismissed. We are accustomed to the separation of scientific, philosophical, ethical, and aesthetic modes of inquiry, and to academic parlance that is verifiable and solemn. Attributed to the great Mark Twain is this apropos comment: 'the researches of many commentators have already thrown much darkness on this subject, and it is probable that if they continue we shall soon know nothing at all about it'.

Our habitual modes of thought and language categories, especially in academia, are inadequate for this synthesis of knowledge, insights, and affectivity. Hyper-rational modes of inquiry are restrictive. Customary intellectual tools that measure, define, analyze, critique, and deconstruct hegemonies have limits. These intellectual processes, while valuable, neither come from nor speak to the depth of human interiority. Hence, they cannot illuminate what is being learned of the comprehensiveness and coherence of the universe. My preferred mentors (Gaston Bachelard, Teilhard, Berry, Heschel) had a great appreciation for many modes of knowing and perceiving, including dreams, stories, imagination, and poiesis. Other sensibilities — such as emotions, intuitions, insights, presentiments, wonder, and wisdom — are valid indicators of knowledge. Interiority and depth perceptions are a place of great vitality, elasticity, and inventiveness.

The human pursuit to understand the cosmos is both an exterior and interior quest. There is an interior aspect to the dynamics and processes, and it is in interiority that we experience the universe. Those who quest to live in a cosmos, experience the cosmos. The intimate immensities of the cosmos are perceived or intuited within these interior expansions of symbolic consciousness of the self. It becomes evident, over time, that this is an encounter. It is possible to learn the data and remain unmoved, but as Bachelard would point out, there has thus been no passionate liaison, no affectivity. The material imagination is not engaged, and only the inferior functions of the intellect, cognition, and rationality are involved.

There are myriad continuities between the breadth of knowledge of the living universe and a depth of inner awareness. This new knowledge expands interiority, and magnifies consciousness. If we can absorb it, our horizons enlarge, our awareness heightens, and our religious sensibilities intensify. Again, Heschel says it best:

> We can never sneer at the stars, mock the dawn or scoff at the totality of being. Sublime grandeur evokes unhesitating, unflinching awe. Away from the immense, cloistered in our own concepts, we may scorn and revile everything. But standing between earth and sky, we are silenced by the sight.[12]

c) How Are We to Live? To Live in a Cosmos

It was in these exterior and interior quests that Berry saw a way forward to respond to the ecological crisis, with a transformed cultural orientation. His intellectual acumen as a historian of religions and culture, and his astute awareness of religious experiences, poiesis, scientific knowledge and so much more became embedded in his dream of the earth and cosmological proposal. This was not for the purpose of expanding knowledge and consciousness. He shaped and interpreted the knowledge to activate dimensions of interiority to respond to the escalating ecological and social crises.

Our cultural and religious maps are not functioning in the interests of a vital biosphere. How are we sure? Look at what is happening to Earth. Look at our economic systems, at the escalating violence, the war on terror, the war on women, consumerism, and billion-dollar arms industries while people starve or have no health care. Anthropogenic climate change is now recognized, but action plans are gridlocked among competing interests and powerful lobbies. Post-truth discourse, fake news, and alternative facts are dulling human sensibilities and shrinking inner and outer horizons of meaning. There is much discussion about why we cannot move effectively on ecological issues.

We cannot perceive an adequate orientation towards the planetary demands of the present. For Berry, responses lie within the cultural visions, social imaginaries or stories. The current versions are dysfunctional in their larger social and ecological dimension, and are not providing direction for a viable future. What stories could give guidance for our era? What gives us an exterior and interior orientation to integrate the most we can know about where we are, and who we

---

12   Heschel, *I Asked for Wonder*, p. 2.

are? The response here is that it is the universe and Earth, in all their complexities, majesty, diversities, and exigencies that educate and orient the deep sensibilities of the human animal. Religion and science need to collaborate to perceive the psychic-spiritual dimensions intimately interwoven in the physical-material. In order to respond to current challenges, we need to appreciate the magnitude and magnificence of existence. To live in a cosmos is to experience it as intimate immensities, which illuminate a path, and radiate radical openness. To live in a cosmos, the emergent universe, as the primary reality, can offer such an orientation. The cosmos is not just 'out there'. It is also within. If we can discover our role in these larger evolutionary processes, there may be hope, as Rachel Carson maintained:

> The more clearly we can focus our attention on the wonders and realities of the universe about us, the less taste we shall have for destruction.[13]

Bibliography

Barad, Karen, *Meeting the Universe Half-Way: Quantum Physics and the Entanglement of Matter and Meaning* (Durham, NC: Duke University Press, 2007), https://doi.org/10.1215/9780822388128

Berry, Thomas, *Dream of The Earth* (San Francisco, CA: Sierra Club Books, 1988).

Carson, Rachel, *Lost Woods: The Discovered Writings of Rachel Carson*, ed. by L. Lear (Boston, MA: Beacon Press, 1998).

Dresner, Samuel, ed. *I Asked for Wonder: A Spiritual Anthology*, Abraham Joshua Heschel (New York, NY: Crossroad, 1997).

Eaton, Heather, ed., *The Intellectual Journey of Thomas Berry: Imagining the Earth Community* (Lanham, MD: Lexington Press, 2014).

— 'The Revolution of Evolution', *Worldviews: Environment, Culture, Religion* 11.1 (2007), 6–31, https://doi.org/10.1163/156853507x173478

Goodenough, Ursula, *The Sacred Depths of Nature* (New York, NY: Oxford University Press, 2000).

Kaufman, Gordon, *In the Beginning: Creativity* (Minneapolis, MN: Fortress, 2004).

Kauffman, Stuart, *Reinventing the Sacred: A New View of Science, Reason and Religion* (New York, NY: Perseus, 2008)

---

13 Rachel Carson, *Lost Woods: The Discovered Writings of Rachel Carson*, ed. by L. Lear (Boston, MA: Beacon Press, 1998), p. 163.

Margulis, Lynn, *The Symbiotic Planet: A New Look at Evolution* (New York, NY: Basic Books, 1998).

Margulis, Lynn, and Dorion Sagan, *Dazzle Gradually: Reflections on the Nature of Nature* (White River Junction, VT: Chelsea Green Publishing, 2007)

Sahtouris, Elisabet, *Earthdance: Living Systems in Evolution* (San José: iUniverse Publication, 2000).

Teilhard de Chardin, Pierre, *The Human Phenomenon*, trans. by Sarah Appleton-Weber (East Sussex, UK: Sussex Academic Press, 1999).

'The Shores of the Cosmic Ocean [Episode 1]', *Cosmos: A Personal Voyage*, PBS, 28 September 1980.

Tucker, Mary Evelyn, and John Grim, eds, *Living Cosmology: Christian Responses to the Journey of the Universe* (Maryknoll, NY: Orbis, 2016).

20. Learning to Weave Earth and Cosmos

Mitchell Thomashow

The living Earth community is a beautiful vision of how to live on this magnificent planet, suggesting that we broaden our perspective to include multiple ways of knowing and being. The fine essays in this volume consider the ethical, spiritual, and perceptual challenges of this vision, offering new approaches to planetary citizenship in an evolving cosmos. This is also an educational challenge. Can we overhaul how we think about learning and teaching so that it is derived from our common aspiration — how to live in a flourishing living Earth community? This challenge encompasses two dimensions. The first is to develop the foundations for this approach to learning. The second is to offer curricular suggestions for multiple educational settings.

Throughout my career I've strived to develop educational approaches to enhance environmental awareness. Two earlier works, *Ecological Identity: Becoming a Reflective Environmentalist* (1995) and *Bringing the Biosphere Home: Learning to Perceive Global Environmental Change* (2001), emphasize the necessity of growing a place-based orientation to ecological awareness, while deepening our understanding of biosphere processes. A new work, *To Know the World: Why Environmental Learning Matters* (2020), applies that philosophy to the challenging issues of our times — migration, the Internet, social justice, networks and connectivity, and adaptation. All these books — indeed, my lifework — concern learning to weave identity and the biosphere. The substrate of Earth and Cosmos is always there. In this short essay, I'd like to briefly explore the educational qualities that are the source of these aspirations. I hope they

 https://doi.org/10.11647/OBP.0186.20

provide readers with an educational foundation for weaving Earth and Cosmos.

The most crucial element in promoting this deepening awareness is in cultivating the ecological imagination, learning how to expand spatial and temporal scale, perceiving the interpenetration of landscape, biosphere, and cosmos, entering the multiple Umwelts of the countless species with whom we share our planetary residency. There are many precedents for this vision. As Andrea Wulf concludes in her magnificent biography of Alexander Von Humboldt 'we can only truly understand nature by using our imagination'.[1] For Humboldt, 'the imagination soothed the deep wounds that reason created.'[2] Humboldt's active imagination was stimulated by his extraordinary observational powers, his ability to synthesize information, his interpretive originality, his dynamic expressive approaches, and their manifestation as exemplified by both his scientific work and his outspoken critiques of colonialism and slavery. What is the twenty-first century version of this sequence of observation, information, interpretation, expression, and manifestation?

I'd like to propose a sequence of learning pathways, ways of knowing that are specifically organized to expand environmental awareness, and hence weave Earth and Cosmos. I describe these learning pathways as 'the five qualities' because they represent distinctive attributes. Each quality entails intrinsic learning processes. All of the qualities and learning processes are simultaneously enfolding and unfolding. They encompass each other while they reveal deepening insights. These qualities are interconnected and mutually reinforcing.

This is not an empirical theory, but rather an informal template, based on four decades of teaching and thinking about environmental learning. I offer this approach in the spirit of educational experimentation and improvisation. There are multiple ways to arrange these qualities. I would like to explain the predispositions that inform my approach.

First, I'm intrigued by the dialectic between perennial and adaptive learning. Perennial learning represents an educational virtue that is consistent across cultural place and time. Environmental insights emerge in similar ways in a variety of cultural settings. Yet, the context of learning

---

1   Andrea Wulf, *The Invention of Nature: Alexander von Humboldt's New World* (New York, NY: Alfred A. Knopf, 2015), p. 336.
2   Ibid., p. 84.

is never the same. People, cultures, and organisms respond to changing circumstances. Hence, learning is also adaptive. In the first decades of the twenty-first century, dynamic environmental change and the acceleration of information technologies are the context for adaptive learning.

Second, I don't think educators spend sufficient time considering how people learn, especially in higher education. Most curricular controversies are substantive. However, *how* you learn is as important as *what* you learn. The skills of lifelong learning are typically internalized when you learn how to learn, and these skills receive insufficient reflective attention.

Third, ecological thinking embodies a paradigmatic shift in how we think about learning. That shift transcends interdisciplinarity per se. It assumes innovative approaches to how we engage as learning organisms in complex environments, how we see ourselves in the biosphere, and how we expand our concepts of place and time.

Fourth, I consider organizational schemes as mandala sand paintings. You create a temporary order of symmetry, coherence, pattern, and meaning, and then you let it all dissipate and recreate it as necessary. Learning is a reflective blend of structure and improvisation, pattern and chaos, coherence and dissonance.

Fifth, the best way to think about any organizational scheme is to personalize it, using it as a way to explore how you learn, how you observe the way others learn, and by considering how learning is a reciprocal relationship between the self, culture, and the environment.

I encourage you to experiment with these qualities and rearrange them to suit your own purposes.

And now for the five qualities.

Observation emphasizes a broadened understanding of biosphere patterns, including the ability to design learning activities and research approaches that enhance perception of global environmental change, an understanding of the relationship between local and global, and the ability to move between spatial and temporal scales. Observation entails perception, identification, and pattern recognition. *Perception* is the development of sensory awareness, so as to apprehend movement, metabolism, pace, and behavior. *Identification* allows an individual to enter the lifeworld (umwelt) of other organisms. *Pattern Recognition* is the ability to assimilate perception and identification by using scale to detect symmetry, cycles, waves, thresholds, interstices, flows, and species interactions.[3]

Information describes the ability to gather data from a variety of sources, organize that data, assess its relevance and application, and understand how to use it effectively. Information entails sourcing, browsing, and networking. *Sourcing* involves understanding the origins of information, its dissemination, its transformation, and how it is manipulated or translated based on opinion and perspective. *Browsing* involves the survey of information, including scanning (seeing the breadth of the field), scaling (understanding its context), focus (knowing how to look more deeply), and granularity (finding its constituent pieces). *Networking* entails mapping information, tracing its routes and paths, determining its speed of transmission (mobility), and understanding who has access to it. An interesting way to conceive of information, and an approach that is facilitated by computer graphics, is the emerging field of information design and visualization. Information design uses the above concepts and develops visualization processes to enhance our understanding of them.

Interpretation is the challenge of generating meaning from observation and information. This includes constructing narrative, amplifying and articulating personal voice, and developing themes and approaches for communicating complex environmental issues. Interpretation entails synthesis, dissonance, and narrative. *Synthesis* is the ability to find coherent relationships within diverse fields of information while finding the essence of ideas and explanations. *Dissonance* reflects the tensions inherent in synthesis, the recognition of nonlinearity, different perspectives, and contrasting possibilities. *Narrative* is the ability to create arcs of unfolding meaning, embodying both synthesis and dissonance through the use of allegory, metaphor, and story. In the twenty-first century, electronic communications make new forms of narrative available and novel forms of expression possible, including the use of diverse media, and reliance on iconography, design, and virtual/visceral matrices, demanding innovative approaches to interpretation.

---

3   For a comprehensive discussion of these issues, and for specific examples from the field of ecology, see Rafe Sagarin and Aníbal Pauchard, *Observation and Ecology: Broadening the Scope of Science to Understand a Complex World* (Washington, DC: Island Press, 2012), https://doi.org/10.5822/978-1-61091-230-3

Expression is the ability to effectively communicate interpretive approaches by cultivating creative possibilities in venues such as storytelling and eloquence, writing and personal reflection, information design and display, artistic mapping, public art, soundscape design, animation and video, music and dance performance, game design, and other forms of iconography and representation. Expression entails imagination, improvisation, and activation. *Imagination* is a unique blend of creativity, visualization, and reflection, allowing the mind to form uninhibited images and possibilities by exploring the unconscious, and melding psyche with the biosphere. *Improvisation* is the ability to spontaneously respond to dynamic changes in the environment by adapting structures of knowledge to new contingencies, or playing with forms and ideas as they emerge. *Activation* is the application of imagination and improvisation through experimentation, innovation, and implementation. Electronic communications enable a spontaneity of response that can have wide (but not necessarily deep) impact in a short period of time. How can expression be simultaneously deep and wide, perennial and adaptive, structured and improvisational, active and reflective?

Manifestation refers to the generosity of interpretation and expression, applying narrative forms to enhance human flourishing in the biosphere. This includes an understanding of social and emotional intelligence, interspecies empathy, the ability to form collaborative connections and challenging learning communities in multiple cultural settings, the ability to engage in creative conflict, and the awareness to improvise in and adapt to diverse learning venues. Manifestation entails generosity, posterity, and flourishing. *Generosity* is the ability to demonstrate kindness, compassion, and respect in service to cultural community and ecosystem integrity. It encourages empathy, dialogue, connectedness, and love. *Posterity* requires awareness of past and future generations, the ability to act with respect for legacy and outcome, and to do so with an expansive time scale. If we combine posterity and empathy, we consider our actions in all of these contexts — intergenerational, multicultural, interspecies, urban/rural, local/global, and cosmopolitan. *Flourishing* is the ultimate goal of environmental learning, to create settings that allow for optimal human thriving in the dynamic biosphere. Flourishing promotes pleasure, virtue, equity, opportunity, collaboration, community, restoration, and reciprocation.

The 'Five Qualities' are integrative pathways for a 'living' education on a 'living' Earth — approaches that enhance wonder, appreciation, gratitude, reciprocity, and service. Reflect again on the meaning of the living Earth community. Humans participate in a complex evolutionary and ecological matrix of species and landscapes. This broad and splendid concept of community demands that we open our minds to the extraordinary circumstances of human awareness in the biosphere, of planetary evolution in the cosmos, of cosmic dimensionality and ethical clarity. Above all, it challenges us to live in a state of wonder and reciprocity, demanding that we live meaningful and purposeful lives. And yet all of this must be made tangible by bringing it home to the places where we live, the habitats that nourish us, our human and more than human neighbors, and how we choose to live complex and fulfilling lives. Now more than ever this is the educational challenge of our times.

Bibliography

Sagarin, Rafe and Anibal Pauchard, *Observation and Ecology: Broadening the Scope of Science to Understand a Complex World* (Washington, DC: Island Press, 2012), https://doi.org/10.5822/978-1-61091-230-3

Thomashow, Mitchell, *Ecological Identity: Beoming a Reflective Environmentalist* (Cambridge, MA: The MIT Press, 1995).

— *Bringing the Biosphere Home: Learning to Perceive Global Environmental Change* (Cambridge, MA: The MIT Press, 2001).

— *To Know the World: Why Environmental Learning Matters* (Cambridge, MA: The MIT Press, 2020).

Wulf, Andrea, *The Invention of Nature: Alexander von Humboldt's New World* (New York, NY: Alfred A. Knopf, 2015).

List of Illustrations

Index

About the team

Alessandra Tosi was the managing editor for this book.

Adèle Kreager performed the copy-editing, proofreading and indexing.

Anna Gatti designed the cover using InDesign. The cover was produced in InDesign using Fontin (titles) and Calibri (text body) fonts.

Luca Baffa typeset the book in InDesign. The text font is Tex Gyre Pagella; the heading font is Californian FB. Luca created all of the editions — paperback, hardback, EPUB, MOBI, PDF, HTML, and XML — the conversion is performed with open source software freely available on our GitHub page (https://github.com/ OpenBookPublishers).

This book need not end here...

Share

All our books — including the one you have just read — are free to access online so that students, researchers and members of the public who can't afford a printed edition will have access to the same ideas. This title will be accessed online by hundreds of readers each month across the globe: why not share the link so that someone you know is one of them?

This book and additional content is available at:

https://doi.org/10.11647/OBP.0186

Customise

Personalise your copy of this book or design new books using OBP and third-party material. Take chapters or whole books from our published list and make a special edition, a new anthology or an illuminating coursepack. Each customised edition will be produced as a paperback and a downloadable PDF.

Find out more at:

https://www.openbookpublishers.com/section/59/1

Like Open Book Publishers

Follow @OpenBookPublish

Read more at the Open Book Publishers BLOG

You may also be interested in:

Earth 2020

An Insider's Guide to a Rapidly Changing Planet

by Philippe Tortell (*ed.*)

https://doi.org/10.11647/OBP.0193

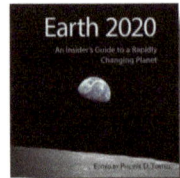

What Works in Conservation 2020

by William J. Sutherland, Lynn V. Dicks, Nancy Ockendon, Silviu O. Petrovan and Rebecca K. Smith (*eds*)

https://doi.org/10.11647/OBP.0191

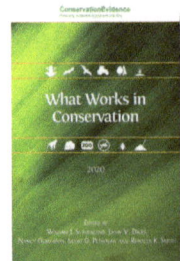

Conservation Biology in Sub-Saharan Africa

by John W. Wilson and Richard B. Primack

https://doi.org/10.11647/OBP.0177

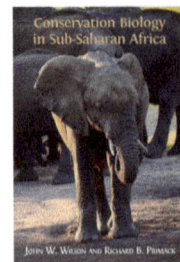

www.ingramcontent.com/pod-product-compliance
Lightning Source LLC
Chambersburg PA
CBHW040147270326
41929CB00025B/3415